Adaptability of the US Engineering and Technical Workforce

PROCEEDINGS OF A WORKSHOP

Kenan Patrick Jarboe and Steve Olson, *Rapporteurs*

Steering Committee on Preparing the Engineering and Technical Workforce for Adaptability and Resilience to Change

NATIONAL ACADEMY OF ENGINEERING

THE NATIONAL ACADEMIES PRESS
Washington, DC
www.nap.edu

THE NATIONAL ACADEMIES PRESS   500 Fifth Street, NW   Washington, DC 20001

The workshop was directly supported by contributions from GE, the US Department of Energy Advanced Manufacturing Office, EY, Allied Minds, Dassault Systèmes, and Nicholas Donofrio. The workshop also relied on past contributions for the *Making Value for America* foundational report from the Robert A. Pritzker Family Foundation, Gordon E. Moore, Cummins, Boeing, IBM, Rockwell Collins, Xerox, Qualcomm, and gifts to the National Academy of Engineering Fund by Jon Rubinstein and Edward Horton. Any opinions, findings, conclusions, or recommendations expressed in this publication do not necessarily reflect the views of any organization or agency that provided support for the project.

International Standard Book Number-13:   978-0-309-47180-0
International Standard Book Number-10:   0-309-47180-X
Digital Object Identifier: https://doi.org/10.17226/25016

Additional copies of this publication are available for sale from the National Academies Press, 500 Fifth Street, NW, Keck 360, Washington, DC 20001; (800) 624-6242 or (202) 334-3313; http://www.nap.edu.

Copyright 2018 by the National Academy of Sciences. All rights reserved.

Printed in the United States of America

Suggested citation: National Academy of Engineering. 2018. *Adaptability of the US Engineering and Technical Workforce: Proceedings of a Workshop*. Washington: The National Academies Press. doi: https://doi.org/10.17226/25016.

The National Academies of
SCIENCES • ENGINEERING • MEDICINE

The **National Academy of Sciences** was established in 1863 by an Act of Congress, signed by President Lincoln, as a private, nongovernmental institution to advise the nation on issues related to science and technology. Members are elected by their peers for outstanding contributions to research. Dr. Marcia McNutt is president.

The **National Academy of Engineering** was established in 1964 under the charter of the National Academy of Sciences to bring the practices of engineering to advising the nation. Members are elected by their peers for extraordinary contributions to engineering. Dr. C. D. Mote, Jr., is president.

The **National Academy of Medicine** (formerly the Institute of Medicine) was established in 1970 under the charter of the National Academy of Sciences to advise the nation on medical and health issues. Members are elected by their peers for distinguished contributions to medicine and health. Dr. Victor J. Dzau is president.

The three Academies work together as the **National Academies of Sciences, Engineering, and Medicine** to provide independent, objective analysis and advice to the nation and conduct other activities to solve complex problems and inform public policy decisions. The National Academies also encourage education and research, recognize outstanding contributions to knowledge, and increase public understanding in matters of science, engineering, and medicine.

Learn more about the National Academies of Sciences, Engineering, and Medicine at **www.nationalacademies.org**.

The National Academies of
SCIENCES • ENGINEERING • MEDICINE

Consensus Study Reports published by the National Academies of Sciences, Engineering, and Medicine document the evidence-based consensus on the study's statement of task by an authoring committee of experts. Reports typically include findings, conclusions, and recommendations based on information gathered by the committee and the committee's deliberations. Each report has been subjected to a rigorous and independent peer-review process and it represents the position of the National Academies on the statement of task.

Proceedings published by the National Academies of Sciences, Engineering, and Medicine chronicle the presentations and discussions at a workshop, symposium, or other event convened by the National Academies. The statements and opinions contained in proceedings are those of the participants and are not endorsed by other participants, the planning committee, or the National Academies.

For information about other products and activities of the National Academies, please visit www.nationalacademies.org/about/whatwedo.

STEERING COMMITTEE ON PREPARING THE ENGINEERING AND TECHNICAL WORKFORCE FOR ADAPTABILITY AND RESILIENCE TO CHANGE

Theresa Kotanchek, *Chair,* Chief Executive Officer and Cofounder, Evolved Analytics LLC
Linda Argote, David M. and Barbara A. Kirr Professor of Organizational Behavior and Theory, Tepper School of Business, Carnegie Mellon University
Ewa A. Bardasz (NAE), President and Cofounder, Zual Associates in Lubrication LLC
Betsy Brand, Executive Director, American Youth Policy Forum
Nicholas M. Donofrio (NAE), Fellow Emeritus and Former Executive Vice President of Innovation and Technology, IBM Corporation
Ann F. McKenna, Director, Polytechnic School, and Professor, Ira A. Fulton Schools of Engineering, Arizona State University
Mary Ann Pacelli, Manager, Workforce Development, Manufacturing Extension Partnership, National Institute of Standards and Technology
Annette Parker, President, South Central College
Wanda K. Reder (NAE), Chief Strategy Officer, S&C Electric Company
Christian D. Schunn, Professor of Psychology and Senior Scientist, Learning Research and Development Center, University of Pittsburgh
Ernest J. Wilson III, Founding Director, Center for Third Space Thinking, University of Southern California

NAE Staff

Proctor P. Reid, Director of Programs
Kenan P. Jarboe, Senior Program Officer
Michael Holzer, Senior Program Assistant

Acknowledgments

This Proceedings of the Workshop on the Adaptability of the Engineering and Technical Workforce was reviewed in draft form by individuals chosen for their diverse perspectives and technical expertise. The purpose of this independent review is to provide candid and critical comments that will assist the National Academies of Sciences, Engineering, and Medicine in making each published proceedings as sound as possible and to ensure that it meets the institutional standards for quality, objectivity, evidence, and responsiveness to the charge. The review comments and draft manuscript remain confidential to protect the integrity of the process.

We thank the following individuals for their review of this proceedings:

Monitor:
Linda P.B. Katehi (NAE), Chancellor Emerita and Distinguished Professor, Department of Electrical and Computer Engineering, University of California, Davis

Reviewers:
Ewa A. Bardasz (NAE), President, Zual Associates in Lubrication, LLC
Keith W. Bird, Chancellor Emeritus, Kentucky Community and Technical College System
Brian Blake, Executive Vice President and Nina Henderson Provost, Drexel University
William B. Bonvillian, Lecturer and Senior Director for Special Projects, Massachusetts Institute of Technology

Venkatesh (Venky) Narayanamurti (NAE), Benjamin Peirce Research Professor of Technology and Public Policy, Harvard University
Gail G. Norris, US Lead Industry Learning Services, Digital Factory/Process Industries & Drives, Siemens Industry, Inc.

Although the reviewers listed above provided many constructive comments and suggestions, they were not asked to endorse the content of the proceedings nor did they see the final draft before its release. The review of this proceedings was overseen by Janet Hunziker, Senior Program Officer with the National Academy of Engineering. She was responsible for making certain that an independent examination of this proceedings was carried out in accordance with standards of the National Academies and that all review comments were carefully considered. Responsibility for the final content rests entirely with the rapporteurs and the National Academies.

In addition to the steering committee, NAE staff, workshop presenters, and session moderators, we thank all the workshop participants for their lively engagement in the discussion and dialogue and for all the ideas generated.

Contents

1 INTRODUCTION AND OVERVIEW   1
Context for the Workshop, 2
Introduction to the Workshop, 4

2 ADAPTABILITY IN AN UNCERTAIN WORLD   5
The Role of Expertise, 6
The Need for Strategy, 7
Placing Multiple Bets, 8
Innovation Derives from Combination, 9
Expect the Unexpected, 9
Speed Is the New IP, 10
A Worldwide Challenge, 10

3 WHY IS ADAPTABILITY IMPORTANT?   12
Preparing for an Uncertain Future, 13
Organizational and Individual Adaptability, 15
The Agile Mind, 18

4 WHAT DOES ADAPTABILITY LOOK LIKE?   21
Individual Adaptability, 22
Interpersonal Adaptability, 24
Team Adaptability, 26
Organizational Adaptability, 29
Changing the System, 31

5   HOW CAN ANALYTICS BE USED TO MAKE
DECISIONS ABOUT ADAPTABILITY?   34
A Data and Innovation Hub, 35
Assessing Adaptive Performance in the Workplace, 37
Measuring Adaptability, 40
Available Measurement Tools, 44

6   TRAINING AND ORGANIZATIONAL CHANGE   45
Career Paths for a Flexible Workforce, 46
Creating a Homegrown Workforce, 47
Aligning Workforce Training with Workforce Needs, 48
A New Normal for Engineering Programs, 49
Evaluating an Adaptability Program, 50
Takeaways and Remaining Questions, 51
Discussion, 52

7   K–12 EDUCATION AND OUT-OF-SCHOOL LEARNING   55
Flex Factor, 56
Made Right Here, 56
P-TECH Schools, 57
Project Lead The Way, 59
Boston After School & Beyond, 60
Key Takeaways, 61

8   POSSIBLE NEXT STEPS SUGGESTED BY
PARTICIPANTS   63

REFERENCES   66

APPENDIXES

A   Workshop Agenda   69
B   Biographies of Speakers and Committee Members   73
C   Participants List   86
D   Suggestions from Interactive Sessions   90
The Nature and Evaluation of Adaptability, 90
Education and Training, 91
Workplace Learning, 92
Building Support for Adaptability, 93

1

Introduction and Overview

In 2015 the National Academy of Engineering (NAE) released the report *Making Value for America: Embracing the Future of Manufacturing, Technology, and Work*. That study observed that technological developments, reengineered operations, and economic forces are transforming the way products and services are conceived, designed, made, distributed, and supported, a transformation that is having profound effects on the world of work. Jobs consisting of repetitive tasks are being disrupted by automation or offshored to lower-cost producers. Workers are being asked to upgrade their skills to become more productive and adaptable. So far these changes have been most notable in manufacturing and high-technology services, but they are poised to also transform enterprises in energy, health care, education, and other sectors.

To explore the effects of these changes on the workforce and on individual workers, the NAE held a workshop November 2–3, 2017, entitled "Preparing the Engineering and Technical Workforce for Adaptability and Resilience to Change." The first goal of the workshop was to increase stakeholders' understanding of both the importance of workforce adaptability and the definition and characteristics of adaptability. The second goal was to provide an opportunity to share best practices for fostering adaptability and to identify needs for future study and development.

The workshop started with a keynote address on the evening of November 2 and consisted of six sessions the following day (the workshop agenda is in appendix A). Biographies of speakers and the workshop organizing committee members are in appendix B. Over 75 participants from a mix of academia, industry, engineering societies, and government engaged in the dialogue (the list of participants is in appendix C).

CONTEXT FOR THE WORKSHOP

As Nick Donofrio, fellow emeritus at the IBM Corporation and chair of the committee that produced *Making Value for America*, observed at the workshop, the object of work is to produce value, whether people are employed in academia, government, or business.[1] But value changes over time, largely because of the influence of technology. Furthermore, value can migrate away from the United States unless US workers can compete with the rest of the world in creating value, said Donofrio. "Like capital in a free market society, value goes to the place where it is most efficiently deployed."

The continually changing demands of the workplace have created an imperative for workers to be adaptable, Donofrio observed. People need to be able to innovate and change in a collaborative, open, and inclusive way. "My biggest fear is that too few people see that binding thread—and that one missed stitch could ruin the entire deal for us. We can't let that happen. That's my passion, that's my energy, that's why I'm here."

Donofrio said that for many years he has preached the value of what's called a *T-shaped educational model*, fostering the development of both broad "soft" skills and deeper knowledge in a particular area. When people are educated very narrowly for a job or to earn a PhD (what he called the *I-shaped model*), they do not have the breadth that enables them to adapt to new circumstances and new challenges. This is true at all levels of education—the high school diploma, 2-year and 4-year degrees, master's and PhD degrees. "The higher education system gets this, for the most part," he said. "But we have to cascade that model down…to K–12 education," in part because many young people will not continue on to higher education.

One way to provide a T-shaped education is through problem-based learning, but such a step requires fundamental changes to educational systems, which need to be more honest with themselves and with students about the skills people will need in the future, Donofrio said. "We don't help people think about this; there's no context for it."

---

[1] Donofrio offered introductory remarks and also participated in a panel and other discussions. His comments from various points in the workshop are consolidated here.

Innovation is happening in places where it might not be expected, he pointed out. For example, the Boy Scouts of America are bringing a more elastic mindset to youth as merit badges are earned for problem solving, not just for the sake of earning merit badges. "There is some evidence that we're on the right path," said Donofrio. But "time is not our friend. Time is our enemy. There isn't enough of it, and we're not going to get any of it back."

An important part of the answer resides at the local level. States can move faster and be more flexible than the federal government. They can establish examples that others can observe and emulate. They also can help make it easier for people to do a job from where they are, not just from where a company is located. "We need to find states that care about this issue," he said; New York, for example, has done an "incredible job of bringing jobs" to the state.

Donofrio also directed attention to a particular part of the workforce: technical workers who do not have a four-year college degree. People who do technical jobs with less than a four-year degree add great value to the economy, he said. "It's a terrible idea that people who graduate with a high school degree or a two-year associate's degree are not skilled enough to do complex data analytics or [work with] big data." People erect barriers to letting them do such tasks, just as countries try to build barriers to protect capital.

As the pace of change in the economy continues to accelerate, these workers need the skills to be adaptable so that they can continue to build the economy, "because you can't build it without them," he said. Society has an obligation to provide these workers with opportunities to "re-school, re-tool, and re-scale" themselves through, for example, a "GI Bill" for displaced workers. If these people do not have the skills to be adaptable, work will flow to the cheapest sector of the workforce and to the highest-educated sector. And as these workers "lose their way, we lose our country," said Donofrio. "Power comes when you have people, but when there is no work, that power is used against you."

Leaders of businesses, educational institutions, and government need to be honest and truthful about what is going on, Donofrio concluded. They also need to be hopeful and point to a better future, because their employees are working hard and need to know that they are working in an environment that supports both honesty and hopefulness.

INTRODUCTION TO THE WORKSHOP

At the heart of *Making Value for America* was "recognition of the impact of globalization, disruptive technologies, and new business models on how things are produced," said Theresa Kotanchek, chief executive officer and cofounder of Evolved Analytics, who chaired the steering committee that organized the workshop. As an example, she pointed to the impact of information technologies.

The digital universe is growing exponentially, Kotanchek said—from 4.4 zettabytes in 2013 to 44 zettabytes in 2020 (a zettabyte equals 1 billion terabytes). Furthermore, emerging markets surpassed mature markets in the generation of data in 2017, and the gap between the two will continue to grow.

She noted that the impact of the Internet of Things is already visible in the digital universe. Data from embedded systems—the sensors and systems that monitor the physical universe—are projected to rise from 2 percent of the digital universe in 2013 to 10 percent by 2020. Advanced sensors, controls, and software applications are connecting the world's machines, fleets, and networks. Advanced analytics are combining the power of physics-based analytics, predictive algorithms, and deep domain expertise. The ability to connect people at work or on the move is supporting more intelligent design, operations, and maintenance, leading to better service quality and safety.

But the tremendous promise of the new technology faces a major obstacle, said Kotanchek. "If we're not enabling people to connect, integrate, and create further value, it will not deliver on its promise." New business opportunities are fundamentally changing the marketplace, the world of work, and every aspect of business operations. "The future belongs to those companies and those people who learn how to take data and convert it to knowledge and information that they can act on." She quoted a statement attributed to Charles Darwin: "It's not the strongest of the species that survives, nor the most intelligent, but the one most responsive to change."

The workshop was designed to tap into the "power of we," she said, by bringing together people from different sectors, including business, education, and government, to share best practices, learn from each other, identify unmet needs, and articulate how to close those gaps.

2

Adaptability in an Uncertain World

> **BOX 2-1**
> **Highlights of Keynote Presentation by Frans Johansson**
>
> - The world is facing an onslaught of trends and developments that are making the future exceedingly uncertain.
> - Organizations and individuals have different strategies for adapting.
> - Individuals need to iterate themselves quickly.
> - Companies need to place multiple bets simultaneously.
> - The best way to prepare for adaptability is by combining building blocks of knowledge.

On February 24, 2011, Airbnb reached its first million bookings, and a year later it had 10 million. On June 12, 2014, Elon Musk released all of Tesla's patents to the world. On January 24, 2017, Amazon became the first streaming company to receive a best picture nomination. What drove these achievements? What made these stories so different from countless failures? The key to answering these questions, said Frans Johansson, Swedish-American writer, entrepreneur, and public speaker, in the keynote address of the workshop, is understanding the forces of innovation and what they mean about the need to adapt.

Johansson has been interested in the nature of adaptability since he was a child. His mother is black and Cherokee, his father Swedish, and he was raised in Sweden, surrounded by reserved, blue-eyed, blond-haired people. Growing up with parents from different cultures, he was exposed to ideas, perspectives, and norms that provided him with a multifaceted understanding of the world. Eventually, he wrote two books on the nature of innovation: *The Medici Effect: What Elephants and Epidemics Can Teach Us about Innovation* (2004) and *The Click Moment: Seizing Opportunity in an Unpredictable World* (2012), where he explored how the features of modern society have changed the pathways and requirements for innovation.

One sign of the increasing speed, spread, and sources of innovation is the decreasing average lifespan of companies, he noted. About half of today's S&P 500 companies will be replaced in the next ten years. The economy has never been more efficient in sorting companies that can adapt from those that cannot, he said. However, just to say that companies need to be adaptable is not enough, because "adaptability is a tricky concept." Organizations have different ways to adapt than do individuals, and nations may adapt in ways different from those of organizations or individuals.

THE ROLE OF EXPERTISE

An oft-repeated assumption is that expertise is the key to successful adaptation. This notion has a long history, said Johansson. For example, Serena Williams, one of the best tennis players in history, has gained her expertise through lifelong training and practice—she says she has never lived a single day without practicing tennis. Her success clearly supports the idea popularized by Malcolm Gladwell (2008) that expertise in a domain can be attained only after practicing for 10,000 hours.

Does the same rule apply in the modern economy? How many hours of practice did Reed Hastings have before he created Netflix, or Richard Branson before running Virgin Atlantic airline? Johansson said that if he were to ask five people in retail about the future of their business, he would get five different answers. "If I get five separate answers when I ask five experts, what does it even mean to be an expert?"

One difference between tennis and retail is that the rules rarely change in tennis. The most recent rule change was in 1961, when it was decided that players were no longer required to keep one foot on the ground during their serve. Today's economy is the exact opposite, observed Johansson: The rules constantly change, often without warning.

In 2007 Nokia was three times larger than the next largest mobile phone competitor. The company "knew the rules of the phone world: Phones were supposed to have cool colors, cool shapes, and cool ring tones." But after the iPhone came out, no one cared about colors or shapes, people wanted cool apps. Within two years, Apple and Google had transformed cellphones, even though their phones were viewed skeptically by experts when they were released. "That's how quickly it could change."

THE NEED FOR STRATEGY

The world is facing an onslaught of trends and developments that are making the future exceedingly uncertain, Johansson observed. Millions of people all around the globe are entering the workforce and trying to figure out the next big thing. Technologies like artificial intelligence and deep learning suddenly have the potential to reshape the workplace and society. Social networks and technology greatly magnify the spread of innovations, as demonstrated in a tweet from Airbnb's founders: "Marriott wants to add 30,000 rooms this year. We will add that in the next two weeks." Changes are behavioral as well as technological. "Parents always said, 'Don't get into a stranger's car.' Now everybody does it" with Lyft and Uber.

Money, access, and resources are available to fund and fuel the growth of virtually any idea. Success is unexpected, making it difficult to understand where to go next. Trends alone do not provide the key, because there are too many forces interacting with each other.

"Let me state this more provocatively," Johansson continued. "What is the point of planning anything or strategizing if the world keeps changing as quickly as it does?" Strategy is necessary because people do not act randomly. Humans always have a rationale for their behaviors. "The purpose of strategy cannot be to figure out the right answer, because you're likely to be wrong," he explained. "It's to convince yourself to act, to move, to execute, to test something. If we're going to understand the heart of adaptability, we have to understand this piece."

The need to have a strategy has four implications for what a strategy should contain, Johansson said: placing multiple bets; combining expertise and skills with new fields, industries, and cultures; preparing for unpredictability; and enhancing the speed of innovation.

PLACING MULTIPLE BETS

First, if developments in the world are unexpected, one must assume that one cannot know all the answers. Therefore, the more bets an organization or individual places, the better.

Companies and individuals must place bets in different ways, Johansson continued. For the individual, bets are determined by how quickly one can iterate oneself to become a new, better version given the surrounding environment. In business, bets set a company up for a number of possible futures instead of just one.

The idea that people who change the world try far more ideas is a prominent finding from research on innovation and applies across the board—from entrepreneurs to artists to scientists to corporations. Picasso created over 50,000 works of art. Einstein published more than 240 papers. Both continually tried new things, unsure of what would succeed. More recently and at the corporate level, Angry Birds is one of the most downloaded games of all time and changed the way gaming companies thought about the possibilities of gaming. But Rovio, the Finnish developers of the game, made 51 games before Angry Birds. This experience enabled them to create a game maximally appealing to the public and with a focused, scalable marketing strategy.

The trend toward smaller and faster bets will accelerate, Johansson predicted. Today, many companies are being funded on a small and incremental basis. People have been taught that the rules of success tend to be rigid, but as the world becomes more unexpected, the opposite is true. Success is based on flexibility, creativity, and speed. Amazon is a prominent example, expanding from a book-selling business to a huge corporation, with a diverse portfolio of departments and investments, that constantly innovates—with movies, Alexa, drone delivery, grocery shopping.... The company places bets as far as it can reach as part of its strategy to remain a dominant commercial power.

The idea of placing multiple bets is different for an individual than for a company, Johansson noted. An individual cannot place multiple bets simultaneously. Rather, the individual needs to determine, "How quickly can I become a new version of myself, given the realities of what's going on around me?"

INNOVATION DERIVES FROM COMBINATION

The second implication Johannsson cited is that innovation comes from combining existing expertise with new fields, industries, and cultures. The extent to which an individual or organization can acquire, combine, and build on new skills determines adaptability.

He cited the example of a hospital in the United Kingdom that was having trouble transferring patients from surgical units to the intensive care unit. The transfer involved two separate teams, creating opportunities for potentially fatal errors in communication. For insights, the hospital decided to study racecar pit crew operations developed by the McLaren organization. New procedures allowed the hospital to develop best practices for speed, efficiency, and communication, resulting in fewer errors.

The best preparation for adaptability is to combine building blocks of knowledge, Johansson said. Education helps inform decisions by providing the tools to acquire new skills. School is not the only venue for learning new things, but it can provide the structure and models to enable the acquisition of new skills.

The United States, China, and Russia have advantages of large populations, a national language, and insularity at the early levels of scaling. But these nations also boast immense diversity. When Chinese nationals return from the United States, their experiences allow new insights and accelerate innovation in their home country. The combination of insights across cultures, fields, and industries inspires innovation in ways similar to the combination of building blocks of knowledge.

EXPECT THE UNEXPECTED

The third implication of a rapidly changing world is that people and organizations need to pay more attention to the unexpected. This implication is in contrast to what happens in schools. Schools are becoming more and more similar in order to remain competitive, but this model is not universally suitable. The endless comparison of statistics and scores in education can stifle innovation, even though the purported purpose of tests is to promote maximal success.

To promote student passion and drive, universities and grade schools need to rethink and revise old and rigid approaches and models, Johansson said. Schools need to focus on the issues students care about rather than focusing on perfection. Successful companies are not those that do everything perfectly but those that have the drive to push on.

Much in life is based on the idea of a predictable path. In school, one grade follows the other. A student who is a grade level above his or her peers is applauded. But after graduation, there are no longer such differentiators.

"If we want to prepare ourselves for the unexpected, for the unpredictable, then we have to be willing to acknowledge that we gain insights, we gain our passions, and we get our knowledge in ways that don't follow a necessarily normalized path," said Johansson. The resulting deeper understanding of the world may cause one to examine and question the unexpected, the outlier, whereas schools actively teach that the outlier is to be ignored. Surprise is a leading indicator of innovation.

SPEED IS THE NEW IP

The final implication Johansson cited concerns intellectual property (IP). With the massive changes that outliers can produce, speed becomes more important than IP. IP matters only as long as an idea or a product is relevant, but the shelf life for relevance is shrinking. When Elon Musk released Tesla's patents to the world he dramatically accelerated the speed of innovation in that field. The velocity of innovation must match the speed of the changing world, Johansson explained. "You want to be innovating so quickly that you can invalidate your prior patents."

This implication holds for both individuals and organizations. Adaptability depends more on speed than on the knowledge one controls. "The world is changing faster than ever, and it follows that you have to as well."

A WORLDWIDE CHALLENGE

All countries are struggling with these issues, not just the United States, Johansson concluded. Even as the United States worries whether its students are acquiring enough basic skills, other countries are trying to make their educational systems more like the US system to emulate the way it fosters creativity and innovation.

Johansson was optimistic about the United States' prospects in a world of accelerating and unexpected change. "We have a good shot at getting this right," he said. "We have a good chance of getting ourselves into a position where we, as a country, will keep on taking what's good about what we do—the creativity part, the innovation part—and push farther."

"The world is increasingly moving in unexpected ways," he said. "We should move with it."

3

Why Is Adaptability Important?

BOX 3-1
Highlights of Panel Presentations

- Core cross-functional skills that are transferable across multiple industries and functions are critical in a rapidly changing economy. (Guy Berger)
- Workers need both durability (the ability to survive change) and adaptability (the ability to shift to a new direction). (Berger)
- Greater adaptability of workers could raise productivity in the United States by $30 billion. (Berger)
- Market forces and the educational system are pushing people to acquire narrow skills rather than broad skills needed for adaptability. (Berger)
- People need some level of economic security to be adaptable. (Berger)
- Individuals and organizations need adaptability, curiosity, humility, a capacity to innovate and learn continuously, and the ability to deliver results while monitoring trends in industry and technology. (Greg Dudkin)
- Long-term success requires establishing the right culture. (Dudkin)
- Strong leadership and identification of a value proposition can create a culture of adaptability. (Dudkin)
- We are at an inflection point between a past where humans were conditioned to act more like machines and a future where machines will increasingly have the capabilities to be more like humans. (Robert Johnson)

> **BOX 3-1 Continued**
>
> - Colleges need to educate young people for jobs that do not yet exist using technologies that have not yet been created to solve problems that have not yet been identified. (Johnson)
> - Workers have shifted from generalists to specialists to hyperspecialists to neogeneralists. (Johnson)

The first workshop panel, moderated by Wanda Reder, chief strategy officer at S&C Electric Company, looked at the broad question of why adaptability is important. The panelists had very different backgrounds—one was an economist with a technology company, one was the leader of a utility company, and one was a university administrator.[1] Yet they agreed that people working in the modern economy need to be adaptable to deal with an uncertain future.

PREPARING FOR AN UNCERTAIN FUTURE

Productivity growth in the United States is the lowest it has been in 30 years, reported Guy Berger, chief economist with LinkedIn. In manufacturing, labor productivity has not changed for six years, "which is shocking," he said. "Economists are still puzzling about what exactly is going on."

In fact, productivity growth is sluggish worldwide, not just in the United States. Part of it, Berger speculated, may have to do with employees not being certain that their jobs will survive economic downturns, thereby changing the social contract between employers and employees. Whatever the reason, the consequences are substantial. According to a report prepared by PwC for LinkedIn, greater adaptability of workers could create $130 billion of additional productivity annually for the 11 countries examined in the report, including $30 billion in the United States (PwC 2014).

---

[1] Nick Donofrio, IBM Fellow Emeritus, was also a panelist. His remarks from this and other points in the workshop are consolidated in chapter 1.

The trend of low productivity growth could change, Berger acknowledged. The introduction of computers into the economy, or before that electricity, took time to boost productivity growth. Although robotics and artificial intelligence have not yet had a major influence on productivity, they could do so quickly in the future. But the future is very hard to foresee, Berger pointed out. "This is going to sound harsh, but as somebody who used to work in forecasting, [I'd say that] if someone says they know what skills are going to be needed in 10, 25, 50 years, I'd like to sell you oceanfront property in Kansas. There's a lot of uncertainty."

In the face of such uncertainty, durability and adaptability are critical skills for workers, Berger said. Durability is the idea that a worker can survive future change. Adaptability is the idea that employees can shift part of their human capital to a new direction if a change occurs.

The global educational system, however, is generally not preparing people for such a future, according to Berger. Market forces and the educational system are pushing people into acquiring less adaptable human capital. Instead of broad skills in areas such as chemistry, history, or chemical engineering, college students are acquiring relatively narrow skills, such as learning how to use popular programming languages. As graduates become more specialized, they are more vulnerable to change.

But employers have incentives to train workers in specialized rather than generalizable skills—so that, for example, employees may be less tempted to take their skills elsewhere. And employees have an incentive to learn specific skills that make them money right away rather than broader skills that may not pay off until the long term. Government has a role to play in blunting these incentives so that the workforce does not become increasingly specialized and employees gain the types of skills they will need in the long run, Berger said.

In some parts of the economy, particular skills are losing currency. For example, technological change has reduced the need for certain skills in the publishing industry, such as page layout. Also, some skills suddenly become popular but soon fade, such as proficiency with a particular programming language. "A worker who's acquiring [such a skill] will quickly have to acquire another one in a year or two," said Berger. The better option is an education that enables a worker to learn a variety of skills and be ready to move from one to another.

Distinctions between durable and nondurable or adaptable and nonadaptable human capital are actually on a spectrum. Certain kinds of human skills are always going to be needed and might become even more important while others go out of fashion. For example, demands for Twitter-related skills have fallen markedly, while the demand for skills related to other social networks has grown. Yet people with Twitter-related skills have not experienced an obvious economic decline because they have pivoted into other professional areas. This happens throughout the economy. For instance, when the price of oil crashed and many workers employed in the fossil fuel industry in the United States lost their jobs, many appear to have found employment in other areas such as construction. "When you have sets of human capital that potentially can be used in multiple places, people do change."

Core skills that are transferable across multiple industries and functions are critical in the face of uncertainty, Berger pointed out. "I don't like the term *soft skill* because I think it's pejorative," he said, but skills such as leadership, adaptability, or business strategy acumen fall into this category. "That's where I would urge us to dedicate a lot of our efforts." Since the future is uncertain, the best option is to prepare workers for as many futures as possible.

To take advantage of opportunities, people also need geographic mobility and economic security to make the transition to a new job or industry, Berger added. However, the ability to move is disappearing in some parts of the country where the greatest economic opportunities are being created. "Not being able to afford housing in a place that has a lot of jobs is disrupting adaptability in a big way," he said. "Even places that used to be very affordable are no longer affordable. That's keeping people in relatively lower-value economic paths." Geographic mobility means somehow figuring out an affordable housing solution, he said.

In addition, people need some level of economic security to be adaptable. "You need to make sure that people have some sort of economic safety net—whatever that looks like in this kind of situation—so that they feel they can take the chance to adopt and learn new skills to complement their existing ones." If people need to risk their economic livelihood to become more adaptable, their ability to adapt will be blunted.

ORGANIZATIONAL AND INDIVIDUAL ADAPTABILITY

"If [adaptability] is really important for somebody even in the electric utility industry, then it's got to be important for everybody," said Greg Dudkin, president of PPL Electric Utilities. Utilities have historically been seen as relatively stable industries, but even they are now threatened with change. Greater energy efficiency means that electric loads are either not growing or declining. Consumers are looking to generate their own electricity through solar power or other renewables and sell it back to the grid. "What has worked for 100 years is not going to work for even the next 5 to 10 years," he said. "Companies that adapt to change will do very well; those that don't will not."

Long-term success requires establishing the right culture, said Dudkin. For him, that means a constructive culture that is striving to achieve exceptional results. The goal of PPL Electric Utilities is to be the best utility in the country, he said, which requires a culture that "truly engages the employees so that they can reach their full potential." People need autonomy and the ability to achieve mastery that taps into their intrinsic motivation. They need "a purpose that's bigger than themselves."

When Dudkin came to the company, it tended to be passive, defensive, hierarchical, and conflict avoidant. He sought to instill a culture marked by adaptability as well as curiosity and humility. As an example of curiosity, he cited a problem the company had with birds that would dry their wings on utility towers and short out transmission lines. Protecting every tower from birds would be cost prohibitive, so the company turned to data analytics and found correlations between the structures affected by the problem and proximity to populated areas, water, and forests. "We identified a small number of structures that we treated, and we've reduced the impact on our system."

Humility signals an openness to new ideas. "The opposite is arrogance," observed Dudkin. "Arrogant folks don't do well in my organization, because for me it's a sign that they believe they know everything and aren't open to new ideas." People need to be able to innovate, which is a hard thing to mandate. Dudkin has had success bringing in outside advisors to teach employees how to think about things differently. And he has found that the most innovative teams are the ones that have the most diversity, whether in terms of gender, age, race, or position in the company.

He also emphasized continuous learning, which he described as difficult to instill in an organization. Even as his company has won awards for its procedures, he has sent teams to other companies to learn what they are doing better. Whether considering an engineer who does technical work or a lineman who does physical work, "I'm looking for people who are adaptable, who are continuous learners." Interviews give some indication of whether a person will have an agile mindset, but they are not a perfect predictor. His company therefore gives employees a set of tests after they are hired to try to detect correlations between the test results and the ability to work across functions and within teams.

The leaders of an organization can create a culture that blurs the line between front-line workers and management, he said. In this way, all employees can understand what the visions and goals for a company are and help achieve those ends. They will take the initiative to figure out better ways of moving forward, but this initiative needs to originate with leaders and be driven throughout an organization, Dudkin said.

Individuals who are not adaptable to changing conditions will be more at risk in the future, Dudkin continued. They need the same things that companies need: adaptability, curiosity, humility, a capacity to innovate and learn continuously, and the ability to deliver results while monitoring industry and technology trends. Such people are not necessarily the smartest people in an organization, Dudkin explained, but the ones who find out what is happening elsewhere and are able to build on ideas and work well with others.

People who have been with an organization for 30 years are less likely to have been trained to be creative or work across functional silos, Dudkin observed, so organizations need to train them to be innovative. A large number of people in his company have gone through at least one session in which employees learn how to innovate, "and they're coming out excited. It's starting to build on itself—they're getting success, they're seeing that they're adding value, and they're doing it on their own." Employees across the organization are moving from a fixed to a growth mindset.

Dudkin said that his company is very transparent about what is happening both in the industry and in the company, about what the challenges and the opportunities are. Some people are not interested in changing, he acknowledged, but the company is committed to explaining how the future will be different and how employees can prepare for that future.

When Dudkin started at the company, he knew almost every improvement initiative that was happening, "which is really sad." Now he talks with employees about what is going on and has never heard of some of the initiatives. "That's fantastic. They understand what the vision [and] goals of the organization are, and they're taking initiative upon themselves to figure out better ways of moving forward." The leaders of an organization need to create a culture, and remove the barriers, to make that possible, he said.

THE AGILE MIND

In the past, humans were conditioned to act more like machines, whereas in the future machines will increasingly have the capabilities to be more like humans. At present, modern societies are at an inflection point between the past and the emerging future (figure 3-1), said Robert Johnson, chancellor of the University of Massachusetts Dartmouth, who acknowledged the contributions of Heather McGowan in shaping the ideas he presented at the workshop. At the same time, humans are living longer, and paradigm shifts are coming faster and faster. One hundred years went by between the development of the steam engine and the internal combustion engine. More than half a century passed between the discovery of electricity and the development of the telephone and television. But the Internet is only about 20 years old, the iPhone was introduced only about 10 years ago, "and over the next 10-plus years we're going to see even more change," he said.

These new paradigms have produced successive revolutions in the world of work. Before the 20th century, the emphasis was on physical labor, basic engineering, and skills leading to a stable job. From the start

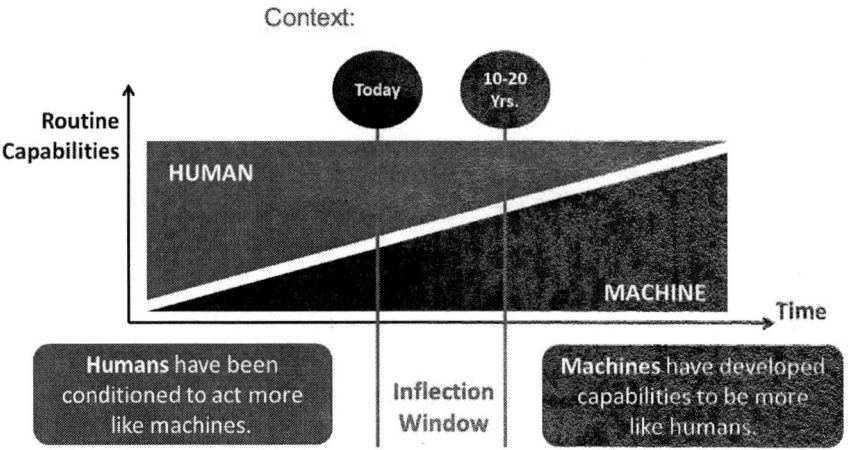

FIGURE 3-1 Society is at an inflection point that marks a transition between humans handling most routine tasks and machines performing the majority of routine tasks. Source: Robert Johnson and Heather E. McGowan, "Future of Work and the Academy," presentation to the National Academy of Engineering, November 2017.

of the 20th century until about 1970, standardization, certainty, and low risks were the hallmarks of the workplace. Since then, deep expertise and disciplinary training have helped workers adjust to the information age. In the future, Johnson predicted, learning agility, adaptability, empathy, and transdisciplinary expertise will be valued. "We have gone from having to be generalists to specialists to hyperspecialists to neogeneralists."

These changes have had profound implications for education. In the past, people went to school to gain knowledge and skills that they could apply to a job over the course of a long working life. As people matured, learned from experience, and gained judgment, they became particularly valuable workers in the middle to later years of their careers. But in today's rapidly changing world, the high-value zone tends to occur earlier in a person's career. With people living longer, perceptions of diminished value tend to extend for longer periods during the later parts of their careers.

Johnson's response to the changing nature of work has been to emphasize the importance of developing an agile mindset. Many of the jobs of the future cannot be foreseen, and many of today's jobs will no longer exist in the same form in the future. People are going to need to be able to recondition themselves for an uncertain future. "The jobs will change, the markets will change, and the skill sets they will need will change," he said. "We are educating young people for jobs that don't exist using technologies that haven't been created to solve problems we haven't identified." Young people need to ask themselves: What is my passion? "If they are passionate," said Johnson, "the productivity will naturally come."

People will need to learn to learn and learn to do throughout their lifetimes. "Value creation will come out of that as a natural process," Johnson said. "We want to give young people knowledge and the power of learning. We want to impart knowledge to them and give them the ability to learn to learn. We want to give them uniquely human skills that robots cannot do. If they have that, they will be adaptable and will be able to create value in the marketplace."

One hallmark of the agile mindset is empathy. Students also need to know how to work collaboratively, solve problems through divergent thinking, be entrepreneurial, and add value. An analogy is the distinction between apps and operating systems. Apps are very specific and designed to work in a particular context. But students need to be educated as "operating systems" so that they can constantly be "upgraded" and work in a variety of contexts. "Do we need majors? I don't know. I'm debating that with myself."

The challenge is particularly great with people who were educated to have a very different mindset, said Johnson. He grew up in Detroit, where people were able to make a good living doing relatively routine jobs. Then, all of a sudden, the jobs were gone. "How do we explain to friends I grew up with in Detroit who went straight into the factory, who are 50-something now, what happened to them? We have to have the courage to say, 'Those jobs are not going to come back, and it's not because they were exported, it's not because of immigrants, it's because the world has changed.'" The nation has to go a step farther, he added, and decide what it is going to do about those changes. "We have to come up with the solutions as a logical next step, because I really do believe that a metalworker, if he or she had the data and understood it,…would figure it out."

People also need to know that it is okay to fail. "If they think they're going to be chastised every time they fail, they're not going to try anything new." That is one of the keys to leadership—creating an environment where there is tolerance for failure.

These changes will make great demands on higher education, Johnson acknowledged. The enrollment of his institution is currently about 9,000 students; a decade in the future, he expects the enrollment to be between 20,000 and 25,000 students. But many of these students will be learning online rather than drawing on the institution's physical infrastructure. "I don't think of our business as brick and mortar anymore. It's click and mortar." To accommodate this more varied learning environment, spaces on campus will need to be open and flexible so that they can be used for many different purposes, whether as maker spaces, faculty offices, or classrooms. The institution can partner with businesses to provide the skills students need, perhaps through online or blended instruction around the edges of an ongoing job. The faculty will need to adopt a more flexible mindset compared with the more fixed mindset that has sufficed in the past. For instance, they might have joint offices, or offices in conjunction with businesses.

In closing, Johnson professed the belief that when the data are put in front of the people in an organization, they can see for themselves why that organization has to change. "It creates a very different dynamic," he said. "It unleashes them."

4

What Does Adaptability Look Like?

> **BOX 4-1**
> **Highlights of Panel Presentations**
>
> - Adaptation can be slow and typically includes a time lag. (Christian Schunn)
> - People adapt whether or not they are aware of environmental changes, but those who are aware of a changing environment tend to adapt more quickly. (Schunn)
> - People who adapt more quickly than others succeed more often in dynamic tasks. (Schunn)
> - Cultural competency, intellectual curiosity, empathy, 360-degree thinking, and adaptability are skills that people need in order to be able to work readily with others. (Ernest Wilson)
> - Adaptation can be affected by societal volatility, uncertainty, complexity, and ambiguity. (Wilson)
> - Team adaptability both shapes and reflects adaptability at other levels, from the individual to the organizational. (Steve Kozlowski)
> - The adaptive capabilities of teams are emergent and can be nurtured and shaped, and interventions can be effective, if underlying mechanisms of team formation and contextual factors are known. (Kozlowski)
> - Adaptability can be either domain general or domain specific. Domain general is about human capital—hiring people with certain capabilities. Domain specific is about training—developing certain skill sets. (Kozlowski)
> - Creating "desirable difficulties" for the learner can be a useful approach to training. (Kozlowski)

> **BOX 4-1 Continued**
>
> - Organizations show considerable variation in the rate at which they learn. (Linda Argote)
> - Four factors—training, transactive memory, effective use of technology, and knowledge transfer—contribute to the different rates at which organizations learn. (Argote)
> - Can adaptability be taught directly or does training need to put people in situations where they are forced to be adaptable in order to learn to be adaptable? (panelists)

The second panel, moderated by Ann McKenna, professor of engineering and director of the Polytechnic School in the Arizona State University Ira A. Fulton Schools of Engineering, looked at the constituent skills and attributes of adaptability at four levels: individual, interpersonal, team, and organizational. The characteristics of adaptability at each level are somewhat different, the panelists observed, but there are commonalities across all four, such as the role of motivation.

INDIVIDUAL ADAPTABILITY

Adaptability is a change in behavior as the environment changes that improves outcomes, said Christian Schunn, senior scientist at the Learning Research and Development Center and professor of psychology, learning sciences and policy, and intelligent systems at the University of Pittsburgh. "We don't talk about adaptability if the environment is not changing, and we don't talk about adaptability if you change in a way that doesn't improve outcomes."

All living things are fundamentally changeable systems, in a process analogous with biological evolution. Even infants learning to crawl or climb adapt to new environments or new constraints. Adaptation can be slow and typically includes a time lag, both when environments change and when they revert to what they were before (Schunn et al. 2001), but people adapt whether or not they are aware of environmental changes. They may "have no idea that things are different, even pretty drastic changes, and yet their body does things differently over time. That's the foundational biological nature of continuous learning."

That said, people who are aware of a changing environment tend to adapt more quickly. "If we are aware of it, we are looking out for change, we are seeing something differently, we shift more quickly." If people have been in a fixed environment for a long time, they typically adapt more slowly than if the environment has changed more recently.

Some people generally adapt more quickly than others and, as a consequence, succeed more often in dynamic and interpersonal tasks (Schunn and Reder 2001). This is true even when experiments control for typical individual differences in cognitive abilities, said Schunn. "The degree to which you tend to adapt quickly drives performance overall to a pretty high extent."

A tradeoff exists between exploiting known resources and exploring the environment. Should a person keep doing the same thing, which has worked well so far, or spend time exploring other ways of doing things that might work less well today but will lead to improvements later? In a completely unchanging environment, the former is the best approach. In a changing environment, exploration allows the development of expertise that may prove useful.

The frontal lobes of the brain are involved in fast adaptations, Schunn explained. When "something is different from what was going on before, that's when the frontal lobes come online, and we force ourselves to do something different from what our muscle memory is telling us to do." Damage to the frontal lobes can severely impair adaptation—as measured, for example, by pattern matching tests—even when people know that they should be adapting (Ridderinkhof et al. 2002). The frontal lobes are the last to develop during childhood and adolescence and are the first to decline with age, though interventions such as even small amounts of exercise can slow this process (Colcombe et al. 2006).

Finally, Schunn observed that motivation is a strong factor in adaptation. People who are constantly looking for ways to improve their performance are more likely to be adaptable (van Vianen et al. 2012). Those who want no more than to prevent a decline in performance tend to be less open to innovations (Hansen 2010). "If somebody comes in with a new idea, it's those who are trying to improve who are going to be receptive to that." This is especially the case for people who believe that they can take an idea and improve upon it, as opposed to people who are so humble that they believe they could not do better. "They have control over change, are curious about change, and are confident in their changes."

INTERPERSONAL ADAPTABILITY

Ernest Wilson, founding director of the Center for Third Space Thinking and professor at the Annenberg School for Communication at the University of Southern California, placed the idea of adaptability in a broader context of interpersonal and interactive skills, thereby providing a "liaison" between individual adaptability and team adaptability. Adaptability is essential in an environment marked by volatility, uncertainty, complexity, and ambiguity. It requires attention to positive opportunities as well as negative risks; it also requires that changes be anticipated and not just reacted to. It should be distinguished from responsiveness, flexibility, resilience, and agility.

Adaptability is driven by one's own initiative and not externally imposed by the demands of one's organization or supervisor, or by an external unexpected event, Wilson said. In some cases, it requires significant changes in behaviors, cognitive skills, and attitudes; in other cases it requires just minor adjustments. In addition, investments in human capital can alter an employee's capacity to contribute to an organization's medium- and long-term success.

Wilson has been involved in a five-year research project on five skills: cultural competence, intellectual curiosity, empathy, 360-degree thinking, and adaptability. Each of the first four skills interacts with and can be leveraged to enhance adaptability. Together, these constitute an educational space that is not that of engineering or business but instead "a third space," he said, "a different way of thinking about the world."

He defined cultural competence as the ability to think, act, and move across multiple boundaries and act effectively. It could be a different country, organization, or unit in the same company or institution. For example, marketing and legal departments have different cultures with their own vocabularies, norms, and appropriate behaviors toward managers, peers, and subordinates. Substantial and sustainable adaptability is impossible if one is ignorant of the culture in which one seeks to adapt, he said. "Cultural competence and adaptability can be seen as mutually reinforcing."

Intellectual curiosity is represented by a hunger to learn and acquire new knowledge for its own sake. It probably cannot be taught from scratch, but everyone can improve their level of intellectual curiosity and their willingness to risk and experiment in order to learn. "The successful adaptor will be more curious about pretty much every aspect of her own job as well as positions that are adjacent to hers up and down the value chain and with coworkers," said Wilson.

Empathy denotes the ability to see the world not only from one's own perspective but from the perspective of others in or outside an organization. This ability to look beyond one's own perspective is critical for adaptability. Learning the formal written rules and requirements of organizational culture is essential but insufficient, Wilson clarified: the superior employee is willing, and even eager, to see the world from the perspective of colleagues and customers.

Wilson defined 360-degree thinking as the ability "to see the big picture, to connect the dots, to think holistically." It intersects with adaptability because every position in an organization is part of the larger whole. Successful adaptors will be able to see their relationships with others in the institution, whether in informal or formal settings. They will be able to recognize that changes are taking place and, as Schunn observed, more readily adapt to those changes.

The Center for Third Space Thinking has been teaching these five skills to executives, graduate students, undergraduates, and others. Especially in the early stages of a career, adaptability and intellectual curiosity are important as employees learn about and adapt to the organization in which they are working. Later, 360-degree thinking is an especially important soft skill as people learn more about the organization. Emphasizing the transitions under way in society can help convince people that adaptability will affect not only their company's performance but their individual performance. "That's the motivational component, which is difficult."

He also observed that exercises in training and leadership development are greatly appreciated. "It turns out that fairly straightforward, simple, hands-on team work activities…have a very positive impact on the macro level to explain the urgency and universality of the changes."

Wilson identified several questions and areas that have been overlooked in research on adaptability. For example, "How do we integrate the development of the soft skills over a career, not just for executives but for…all employees at all levels, and then how do we integrate those soft skills with the hard skills that have traditionally been the domain of engineering professionals?" An area that warrants research attention is the need to anticipate changes rather than reacting to them, and another is the need for adaptability to be driven by an individual's desire and intention to improve performance, rather than by the boss saying, "Here are the four things you must do to adapt or you're fired." Finally, research has often not differentiated between adaptations that require significant changes in behaviors, cognitive skills, and attitudes and those that require minor adjustments.

TEAM ADAPTABILITY

Individuals are embedded in social organizations' hierarchical structure, said Steve Kozlowski, professor of organizational psychology at Michigan State University, and that top-down context influences and constrains behavior at lower levels of the system, which can facilitate or inhibit adaptation. In this way, team adaptability both shapes and reflects adaptability at other levels, from the individual to the organizational.

At the same time, team processes produce bottom-up phenomena shaped by the attributes of individuals and how they interact collectively. As teams go thought the process of being formed and emerging, they are malleable. If one understands the mechanisms underlying this process of emergence, the formation of teams can be shaped and influenced, said Kozlowski.

He illustrated with three hypothetical individuals—Alpha, Bravo, and Charlie—who interact over time and with increasing levels of interaction in a particular context to create an organizational configuration (figure 4-1). Understanding the process mechanisms that underlie those interactions, including the mental models that people share, makes it possible to influence outcomes, he explained. The outcome may be a convergence where everyone comes to the same view, or it may be divergent and more distributed, where different experts have different knowledge sets and interact as a network. "If you want to shape those, you want to come in during the process," Kozlowski said. "Once [an outcome] emerges, it has persistence and will tend to influence subsequent interactions."

As the other speakers in the session noted, adaptability encompasses a number of concepts. It's been defined as "performance capabilities [where] changing external contingencies require rapid shifts in role requirements across team members" (Kozlowski et al. 1999, pp. 241–242). Burke et al. (2006, p. 1192) defined it as "an emergent phenomenon that compiles over time from the unfolding of a recursive cycle whereby one or more team members use their resources to functionally change current cognitive or behavioral goal-directed action or structures to meet expected or unexpected demands." Baard et al. (2014, p. 50) defined it as "cognitive, affective, motivational, and behavioral modifica-

WHAT DOES ADAPTABILITY LOOK LIKE? 27

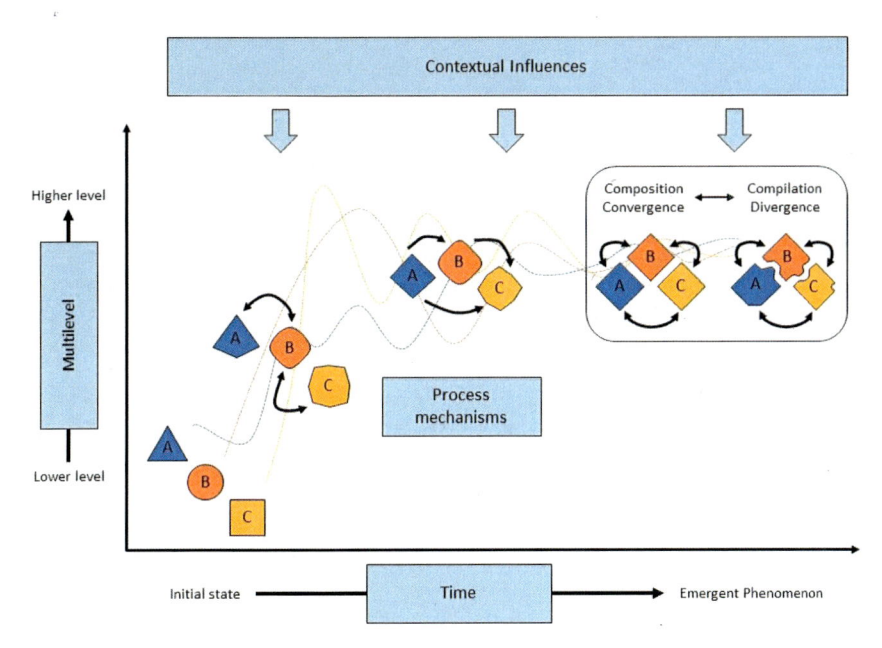

FIGURE 4-1 Interactions among individuals Alpha (A), Bravo (B), and Charlie (C) over time can produce convergent or divergent perspectives. Source: Kozlowski et al. (2013).

tions made in response to the demands of a new or changing environment, or situational demands."

Adaptability can be incremental, continuous, and forecastable, or it can be unpredictable and disruptive. Modifications may be cognitive, affective, motivational, or behavioral. If role requirements change, the performance capabilities of the people on a team need to be flexibly redeployed, Kozlowski said, and this process can involve many forms of adaptation.

The research literature tends to divide such adaptations into two categories (Baard et al. 2014) (figure 4-2). One, which Kozlowski labeled "domain general," considers the broad characteristics of individuals and teams that are predictive of adaptation. This approach looks at desirable performance capabilities such as dealing with emergencies, handling stress, managing uncertainty, learning, and being culturally, interpersonally, and physically adaptive. It then examines the kinds of general cognitive, ability, and personality characteristics that are broadly predictive of those kinds of capabilities.

Performance Adaptation Taxonomy

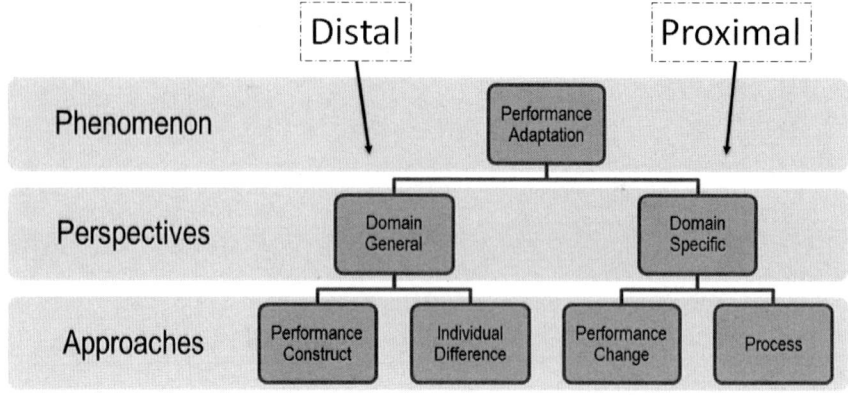

FIGURE 4-2 Taxonomy of types of adaptability. Source: Baard et al. (2014).

The other, labeled "domain specific," is more about particular kinds of expertise in a particular context. It focuses on how individuals and teams operate in a particular situation requiring response to a change. It requires a diagnosis of the situation and development of potential solutions.

The domain-general approach is more selection oriented, whereas domain specific is more process oriented. Domain general is a more distant kind of predictor, and domain specific is more proximal. Domain general is about human capital—hiring people with certain capabilities; domain specific is about training—developing certain skill sets.

From a training and development viewpoint, it is important to remember that team adaptability is emergent and evolves over time. This allows for teams to think about and prepare a repertoire of possible responses for future events before a course is locked in, including developing a depth of expertise and a repertoire of skills for specific situations.

This team dynamic is influenced by the complexity of the task facing the team. Tasks vary in how much load they put on individuals or teams and on the nature of the adaptive process required. When a task imposes a low load, people can develop strategies, set goals, go through a developmental process, and be monitored around the skills needed for the task. When the task load is high, intervention may be necessary to coordinate a team, adjust strategies, and update situations. "You can think about ways that you need to intervene to assist them, and then when it's over you can help lead a team or individuals through a reflection process" to help the team continue to adapt.

Creating "desirable difficulties" for the learner can be a useful approach to training. Education and training are often "proceduralized" so that people learn rules designed to achieve particular goals. People who know proceduralized rules and how to apply them often look as if they are performing at a higher level than people who are exploring and making mistakes. But when things start to change, the people with proceduralized skills know only what they were trained to know and as a result have more difficulty grappling with change, whereas those who are guided to explore adapt more effectively. Individual differences can also be important, whether in terms of goal orientation versus performance orientation or avoidance of mistakes and difficult tasks.

Teamwork typically requires members to regulate attention and effort around multiple goals—individual and team—and to shift the focus of regulation as the situation dictates. This dynamic self-regulation accounts for both individual and team learning and performance (DeShon et al. 2004). Such processes can be modeled computationally, and Kozlowski and his colleagues (2016) are developing such models to look at different configurations of team characteristics and leadership structures and explore what kinds of structures are more robust and resilient and help promote adaptation.

Kozlowski concluded with three takeaway points:

- Individual and team adaptive capabilities are emergent and can be nurtured and shaped.
- Interventions can be effective if underlying mechanisms of team formation and emergence and contextual factors are known.
- Clearly specifying the types of adaptation, the levels of system adaptation involved, and relevant psychosocial mechanisms can generate knowledge that will help inform interventions.

ORGANIZATIONAL ADAPTABILITY

All of the previous levels of adaptation come together at the organizational level. Like individuals and teams, organizations can learn, observed Linda Argote, David M. and Barbara A. Kirr Professor of Organizational Behavior and Theory in the Tepper School of Business at Carnegie Mellon University and director of the Center for Organizational Learning, Innovation and Knowledge. She cited a study of advanced jet manufacturing in which it was shown that the total number of labor hours required to produce each aircraft drops with the cumulative number of aircraft produced, with improvements decreasing over time. This kind of learning curve has been observed in both the manufacturing and service sectors, with performance improvements evidenced by reduced costs, better quality, and enhanced service timeliness (Argote 2013).

Organizations show considerable variation in their rate of learning. Some exhibit dramatic improvement, while others show little or no learning. Even different parts of the same organization that have the same structure and make the same product may learn at different rates and approach different asymptotes (Argote and Epple 1990).

Argote cited four factors that contribute to the different rates at which organizations learn:

- training
- developing transactive memory
- using technology effectively
- transferring knowledge.

The first factor, individual training, improves organizational learning in ways described by previous speakers in the session. It affects individuals in different ways, but organizations can shape training so that it has an optimal effect on organizational learning.

The second factor that contributes to organizational learning is transactive memory, or knowledge of who knows what. The term was coined by a psychologist to describe the specialization in roles and responsibilities that married couples tend to develop (Wegner 1987). Argote explained that, in the workplace, transactive memory allows members of a team to specialize and share capability. "This cognitive specialization makes it easier to coordinate, because you know who is going to do which task, thus saving time and improving performance." In fact, research has demonstrated that providing access to more information than one individual possesses can improve performance (Ren and Argote 2011), and team training that involves observation by the group enables the acquisition of tacit or hard-to-articulate knowledge.

Studies of such varied organizations as software design teams and surgical teams have demonstrated that team members who work together longer perform better and work faster than those with less experience working together. Strong teams match the most qualified members to the tasks they can do best, which promotes problem solving "because you know whom to consult if something unexpected comes up." These sorts of memory systems are especially valuable in dynamic environments, where team members need to consult with and rely on others for advice. Argote mentioned the possibility of groups becoming ossified if they are kept together for a long time, but in most cases enough people move into and out of groups to keep that from happening.

The third factor that can improve organizational learning is the effective use of technology. The introduction of technology can speed progression along the learning curve, through either a sudden improvement (that is, a shift in the intercept of the curve) or a more extended improvement in the rate of learning. Adoption of a new technology is facilitated by strong teams with transactive memory. In addition, individuals and teams can embed expertise in a technology system, which then can spread the knowledge to other parts of an organization (Hwang et al. 2015).

The fourth factor, knowledge transfer, is the process through which one unit of a firm benefits from experience acquired in another (Argote and Ingram 2000). For example, a new procedure developed at one plant can be transferred to another to improve the overall performance of a firm. Argote and her colleagues have seen examples of this knowledge transfer process working across shifts in a manufacturing plant (Epple et al. 1996), products in a manufacturing facility (Egelman et al. 2017), geographically distributed establishments of a financial services firm, and even across units of pizza franchises. Figure 4-3 shows how quickly a second shift can learn from a first shift as both move along a learning curve. Furthermore, understanding how this knowledge transfer works allows learning to be incorporated in the training process.

CHANGING THE SYSTEM

During the question-and-answer session the four panelists considered whether adaptability can be taught directly or whether training needs to put people in situations where they are forced to be adaptable in order to learn to be adaptable.

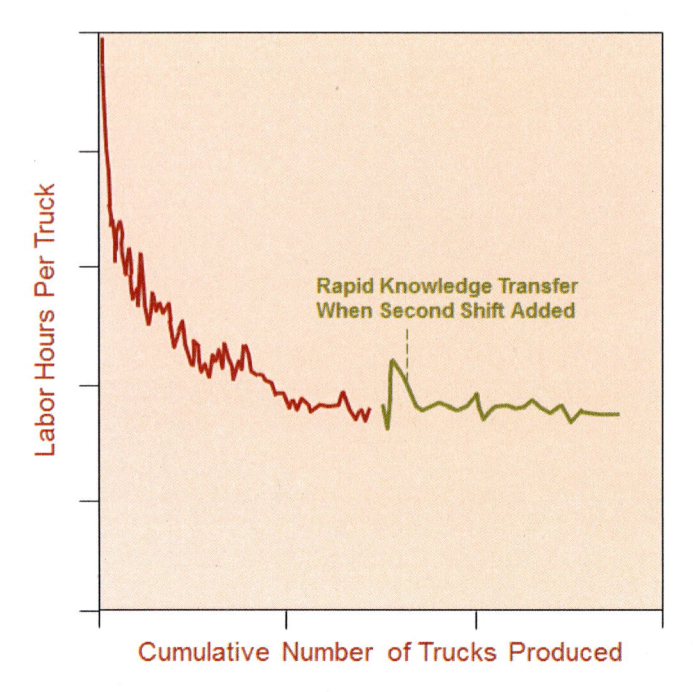

FIGURE 4-3 When a second shift was added in a manufacturing plant, productivity briefly dropped, but it rapidly recovered as knowledge was transferred across shifts. Source: Epple et al. (1996).

In response to that question, Kozlowski elaborated on the difference between domain-general and domain-specific skills. Domain-general skills, such as being open rather than closed to new experiences, can be helpful. But in specific settings, domain-specific knowledge, often gained in the context of teams, can be irreplaceable, although keeping teams together can be expensive and difficult. Education can give people not only the skills specific to particular domains or tasks but also experiences that enable them to be more flexible within a domain, which may be more effective than broader training for adaptability. As he explained, "in a particular setting or environment [that requires] an airline pilot to be able to recover from an emergency, this is about skills flying an airplane, and not just handling emergencies."

No magic bullet can address adaptability, said Wilson; it is one of a set of soft skills that need greater emphasis at all levels of education. The National Academy of Engineering has moved in this direction through such initiatives as the Grand Challenges Scholars Program and the report *The Engineer of 2020: Visions of Engineering in the New Century* (NAE 2004). "Educational institutions need to be more intellectually and professionally aggressive to say that adaptability and other kinds of soft skills are a matter of national performance. It's a big deal. It's not just something that we can tack onto existing things."

Schunn called attention to the risk of "unlearning." If people learn soft skills but then spend 10 years in an environment where they do not use them, they will unlearn many of them. The skills will come back faster if they are needed again, but "you're now going to need some retraining to shake the rust off all those long-forgotten" skills. The better option is to put people in situations where they can use and keep them.

Argote emphasized the importance of learning by doing. "Whether it's projects in school or projects at work, people learn more from [those] and are more likely to be able to adapt their knowledge than from didactic lectures."

Another prominent topic in the discussion session was the continued need for hands-on work as the workplace becomes more automated. Argote maintained that there would still be room for hands-on work, though perhaps not for as many people who worked in those areas as did previously. Schunn agreed, noting that the largest gap in the STEM workforce is not at the bachelor's, master's, or PhD levels but at the technical associate's degree level.

Kozlowski pointed out that automation will have different effects depending on occupations. "A lot of people are going to be displaced by various technological changes." However, many of these people have not been warned that their jobs are at risk and that they will need to adapt. Part of the challenge is educational; part has to do with parenting. "But there are no support systems to try to change those minds."

5

How Can Analytics Be Used to Make Decisions About Adaptability?

BOX 5-1
Highlights of Panel Presentations

- Labor market data can inform a broad set of stakeholders about the best ways to address workforce shortages and skills gaps. (Sandee Joppa)
- Information drawn from online job posting systems across all occupations can help identify the foundational "agility skills" that are in demand. (Joppa)
- There is no single way to measure adaptability, given its many attributes that vary from one job to another; instead, it is embedded across the employee lifecycle. (Tracy Kantrowitz)
- Many jobs require a core set of adaptability skills including creative problem solving, working across teams, and comfort with multidisciplinary environments as employees learn new technologies and procedures. (Kantrowitz)
- The US Army uses Individual Adaptability Theory to train adaptive leaders, involving a small amount of classroom time and a number of practical experiences such as problem solving and teamwork. (Susan Straus)
- Evaluation of the effectiveness of this training indicates that the program is successful but more in-depth evaluation is needed. (Straus)
- Adaptability assessment tools are available but this work is still in its early stages and best practices or best instruments do not yet exist. (Kantrowitz and Straus)

In the workshop's third panel, moderated by Annette Parker, president of South Central College, presenters described methods to analyze and shape adaptability and programs designed to foster adaptability. For example, labor market data can help organizations and individuals understand their environments to better facilitate adaptation, and measures of individual and organizational adaptability can guide decisions and future actions.

A DATA AND INNOVATION HUB

The object of RealTime Talent, explained Executive Director Sandee Joppa, is to enable more informed, market-oriented decisions throughout the Minnesota workforce and education ecosystem by engaging a broad group of stakeholders. The organization works with employers and employer associations in key industries in the state, large higher education institutions—including Minnesota State University and the University of Minnesota—and the state government. It "brings people to the table to [explore] what are we doing about supply and demand, what are we doing about the workforce shortage, how might we make better decisions, and [how can we] take better actions if individual constituencies are informed."

Minnesota has strengths and weaknesses in developing, recruiting, and retaining talent, she said. Employers represent diverse and vibrant industries that provide historically well-paying jobs—Minnesota ranks 13th in the nation in income per capita and 2nd in the Midwest. The overall workforce is well educated and trained, with high participation rates compared with national averages. In the near and long term, demand for workers across skill sets is expected to continue to grow, at 1.5 percent annually across all sectors. However, workforce growth will slow in the near term as current workers age and the number of new workers declines. Net out-migration will prevent the state from augmenting its slow workforce growth—about 6,000 more workers leave the state than migrate in each year. And the impact of employment disparities across racial and ethnic groups will grow as the workforce becomes more diverse, with the population of people of color expected to grow by 50 percent over the next 20 years. Given these trends, the state will have a projected worker gap of 287,000 people by 2022, with about half of that shortage in the Minneapolis–St. Paul area.

To illustrate the innovative approaches that RealTime Talent is bringing to Minnesota's labor market, Joppa reported that the organization has been working with the state legislature to bring a new job posting system to the state. After evaluating 25 online job posting systems, it selected one that algorithmically matches a person's interest with what a job requires. The matching algorithms are blinded to remove bias that might be introduced, for example, by where job seekers went to college, where they worked last, and what their last names are. The state is also seeking to eliminate disparities in employment by raising the labor force participation and employment rates of all racial and ethnic groups to match or exceed those of native-born whites, increase domestic migration to a net positive of 5,000 people per year, and maintain international immigration rather than letting it slow. If these steps are successful, they could reduce the state's worker gap by almost a third, to about 200,000, Joppa said.

RealTime Talent has divided the state into regions, each with its own characteristics. For example, the economy of the northeastern part of the state is based on mining, logging, and paper and apparel manufacturing; average income is low and unemployment is high. The southeastern part of the state is focused on textile and food manufacturing and has low unemployment and a low average income. These regional differences are a key factor in the state's educational system, particularly the Minnesota State system, which offers both two- and four-year degrees. "We are getting employers, educators, and other people together at the table to help solve problems in and create opportunity in those geographies," Joppa reported.

A popular product of RealTime Talent's work, according to Joppa, is its one-page summaries of job data in a region. In the Twin Cities area, for example, the jobs most in demand are registered nurse, customer service representative, administrative assistant, project manager, and business analyst.

This analysis also yields information on the top ten foundational "agility skills" that are in demand (figure 5-1). Skills shown in a lighter color in the right-hand column are specific to engineering/technical occupations compared to all occupations.

These kinds of data can help people take advantage of workplace opportunities, said Joppa. "We are always on the lookout for a new data source, a new perspective, or a new technology that we can bring to the state."

Foundational "Agility Skills" in demand

All Occupations
1. Oral & Written Communication Skills
2. Attention to Detail
3. Integrity
4. Problem Solving
5. Customer Service
6. Creativity
7. Work Independently
8. Organizational Skills
9. Leadership Skills
10. Team Oriented

Engineering/QA/Technical Occupations
1. Oral & Written Communication Skills
2. Problem Solving
3. Project Management
4. Troubleshooting
5. Attention to Detail
6. Creativity
7. Self Starting/Self Motivated
8. Integrity
9. Work Independently
10. Management Skills

RealTime Talent

FIGURE 5-1 Lists of skills in highest demand by Minnesota employers. Source: RealTime Talent, based on aggregate job posting data from TalentNeuron Recruit (www.wantedanalytics.com), accessed 10/18/2017.

ASSESSING ADAPTIVE PERFORMANCE IN THE WORKPLACE

The changing work environment is increasingly demanding an adaptive workforce, observed Tracy Kantrowitz, director of talent solutions at PDRI. She reported the following indications that change is frequent in organizations and increasingly requires employees to adapt to new situations:

- Employees have greater interdependence and now work with an average of ten people to get a job done.
- Organizations change frequently, with the average employee experiencing some form of organizational change—such as a change in leadership, a merger or acquisition, or a restructuring—every seven months.
- Increasing numbers of jobs are knowledge based, with 82 percent of employees doing work that requires analysis and judgment.
- Companies are often geographically dispersed—the amount of work done with coworkers in another geographic location has increased 57 percent in just the past three years.

- The demographic profile of employees is changing; as baby boomers retire and a new generation enters the workforce, employee work preferences are changing.

Adaptability is a multidimensional concept that varies from one job to another. An air traffic controller may have to deal with crisis or emergency situations on the basis of real-time information. An executive assistant may need to handle unpredictable or uncertain circumstances. Engineers need to keep abreast of new tasks and technologies and procedures for working and solving problems creatively. "These all call for different kinds of adaptability," said Kantrowitz.

In addition, many employees are encountering new and intensified attributes of the modern work environment. They need to handle work stress as companies strive to do more with less. "Do they remain composed or do they let that derail their performance?" asked Kantrowitz. "This is a type of adaptability."

Many jobs require creative problem solving, working across teams, and a level of comfort in working with multidisciplinary teams to arrive at breakthrough innovations. Employees also need to learn new technologies and procedures. "How do people stay abreast of new knowledge and methods once they complete their formal education? How do they adapt to new information? How do they continue to grow and develop in their careers?"

Interpersonal adaptability in how people communicate, approach others, and tailor their messaging to different stakeholders is a critical feature of many jobs. Cultural adaptability ensures that people are able to work as a team with people who are from different backgrounds and have different values. Some employees need physical adaptability to work in certain jobs.

Given the many forms of adaptability, there is no single way to measure this attribute, said Kantrowitz. Instead, it needs to be measured across the employee lifecycle, from the determination of adaptive performance requirements to selection of more adaptable employees to management of adaptive performance.

Kantrowitz described the Job Adaptability Inventory (JAI) as a method to determine the adaptive performance requirements of a job (Pulakos et al. 2000) and identify which dimensions of adaptability are most relevant. The results are used to determine the selection of an appropriate individual assessment to identify which people may be more predisposed or more likely to perform well in situations that require

HOW CAN ANALYTICS BE USED? 39

adaptability. The result is a more holistic picture of adaptability rather than a single measure.

To illustrate application of the Job Adaptability Inventory, Kantrowitz picked four jobs and analyzed how the adaptability requirements differ among them (figure 5-2). Assessments of cognitive aptitude are based on abilities such as solving problems creatively, learning new tasks and procedures, and coping with uncertain and unpredictable work situations. Assessment of noncognitive traits is useful in identifying individuals who are more likely to perform well in jobs that require interpersonal adaptability, cultural adaptability, and handling work stress. And assessment of physical adaptability might look at a person's capacity to handle emergencies and physical tasks.

Other measures of adaptability include past experience with adapting, interest in adaptive situations, and self-efficacy to adapt (Pulakos et al. 2002). These measures result in significant prediction beyond cogni-

	Research Scientist	Engineer Support	Police	Installation/Repair
Dealing with Emergency or Crisis Situations			High criticality	Critical
Handling Work Stress	High criticality	High criticality	High criticality	Critical
Solving Problems Creatively	High criticality		Critical	
Dealing with Unpredictable Work Situations	High criticality	Critical	High criticality	Critical
Learning Work Tasks and Technologies	High criticality		High criticality	
Demonstrating Interpersonal Adaptability	High criticality	Critical	High criticality	Critical
Demonstrating Cultural Adaptability			Critical	
Demonstrating Physically Oriented Adaptability			High criticality	Critical

Pulakos et al. (2000)   ● High criticality   ● Critical

FIGURE 5-2 Sample assessment using the Job Adaptability Inventory, showing that adaptability requirements vary by job. Source: Pulakos et al. (2000).

tive ability and personality, and past experience is the best predictor, said Kantrowitz.

She also pointed out that companies are often interested in whether teams can be adaptive, in which case they may use individual adaptability measures and team variables to forecast how well the teams are likely to perform adaptively (Pulakos et al. 2015).

Finally, Kantrowitz noted the importance of training adaptable leaders (Mueller-Hanson et al. 2005). Training should incorporate many opportunities for emerging leaders to be exposed to situations requiring adaptability so they have a catalogue of experiences on which to draw. An iterative process of practice, feedback, and practice is necessary.

MEASURING ADAPTABILITY

Susan Straus, senior behavioral scientist at RAND, described a RAND evaluation of a course that the US Army developed to train adaptive leaders (Straus et al. 2014). The ten-day US Army Asymmetric Warfare Adaptive Leader Program (AWALP) was based both on the Individual Adaptability Theory (I-ADAPT; developed by Pulakos et al. 2000) and on outcomes-based training and evaluation. It was for noncommissioned officers and junior-level commissioned officers, with low student-instructor ratios thanks to small groups of students and larger groups of instructors (called *guides*). Unlike most army training, which is classroom based, this course had a small amount of classroom time and a number of practical experiences, such as problem solving and teamwork.

As mentioned earlier by Kantrowitz, the I-ADAPT approach has eight dimensions of adaptability, characterized as core, supporting, or enabling:

- handling crisis and emergency situations (core)
- handling stress (core)
- thinking creatively (core)
- dealing with changing or ambiguous situations (core)
- interpersonal adaptability (supporting)
- cultural adaptability (supporting)
- physical adaptability (enabling)
- learning tasks, technologies, and procedures (enabling).

Most of these dimensions are intangible, Straus pointed out, which means it can be difficult to assess them. Self-report and observational measures can be time consuming and are subject to various biases and threats to validity. "Many leaders already think they are adaptable," she explained, "so if you ask them how adaptable they are, they will say they are very adaptable." Thus they have little room to improve on self-reports, and courses tend to show little change.

The evaluation used multiple measures and methods (described in Alvarez et al. 2004). For example, most course evaluations are done with an end-of-course survey asking participants what they got out of the course, but "what people think of the course and whether they learn are two different things," Straus clarified. The course evaluation measured changes in knowledge, attitudes, and behavior and whether learning was transferred to performance that made a difference on the job.

An objective measure of transfer is very difficult, according to Straus; the measure used was one of perceived rather than objective transfer. Surveys measuring AWALP students' reactions to the course used both closed and open-ended questions. Examples of the former included the following statements:

- AWALP guides were knowledgeable about the subject matter.
- AWALP guides effectively facilitated after-action reviews and group discussions.
- The feedback I received from AWALP guides enhanced my learning.
- Course materials supported the learning objectives.

Examples of open-ended questions were:

- What did you like best about AWALP?
- What aspects of AWALP should be changed? How would you change them?
- Will AWALP change the way you lead others? If so, how?

Changes in the learners' declarative knowledge were measured using multiple choice pre- and posttraining tests. For example, one of the 30 questions was:

The best definition of adaptability is:

a. Having the capability to complete something in a different way than you have in the past
b. An effective change in behavior in response to an altered situation
c. Constantly changing to keep the enemy off balance
d. Being able to effectively respond to crisis or emergency situations

Changes in the learners' attitudes about adaptability were measured using a pre- and posttraining survey and a multidimensional approach, with items assessing students' experience and need to be adaptable on the job, interest in engaging in adaptive performance behaviors on the job, and self-efficacy for the behaviors. The survey focused on six of the eight dimensions of adaptability emphasized in the course: thinking creatively; interpersonal adaptability; cultural adaptability; learning tasks, technologies, and procedures; handling ambiguity; and decision making under stress. Questions were based, in part, on Pulakos et al. (2000, 2002) and Ployhart and Bliese (2006). Figure 5-3 shows a question about dealing with ambiguous situations.

The survey also included questions about putting adaptive performance behaviors into practice in terms of managing others (based on White et al. 2005). Behavioral learning was measured with student and guide ratings of team performance in several practical exercises. Raters evaluated the degree to which the exercises required different dimensions of adaptive performance and how effectively the teams performed the behaviors. Finally, the evaluation collected data on individual differences that are likely to predict adaptability, such as openness to experience, learning goal orientations, and motivation for training.

The perceived transfer of training was assessed by talking to the graduates three and six months after they completed the course. Evaluators also sought to talk with the students' supervisors, which was challenging because many supervisors did not respond to requests for a discussion or the students wanted to protect the supervisor's time and did not respond to requests for their supervisor's contact information.

Item	Experience	Interest	Effectiveness
		1 = disagree strongly, 2 = disagree, 3 = disagree somewhat 4 = agree somewhat, 5 = agree, 6 = agree strongly	
1. Deal with a situation where things are not "black and white"	☐ Never ☐ A few times/year ☐ Monthly ☐ Weekly ☐ Daily	1 2 3 4 5 6	1 2 3 4 5 6

FIGURE 5-3 Sample survey question probing respondents' experiences in dealing with ambiguous situations. Source: Susan G. Straus, "Measuring Adaptability," RAND, presentation at the National Academy of Engineering workshop, November 2017.

"That may be more specific to the army than to other organizations, but it was a definite challenge," said Straus. Examples of the questions were:

Have AWALP graduates changed professionally as a result of training? Do they do any of the following more than they did before the training?

- Mentor and train subordinates
- Delegate tasks
- Seek input
- Conduct after-action reviews
- Brief commander or senior leader

For graduates, have your attitudes about AWALP changed 3 months and 6 months postgraduation?

- Recommend to others
- Recommended course changes
- Challenges in applying concepts

The evaluation showed a convergence of results providing strong evidence for the success of the training, said Straus. The students had very favorable attitudes toward AWALP, which were sustained over time. They showed large increases in declarative knowledge about adaptability, much higher self-efficacy at the end of the course than at the beginning, greater interest in being adaptable, improvements in judging the need for adaptability, and, according to both supervisors' and self-reports, greater awareness of the applicability of the course's principles on the job. Straus noted that the question of how to measure the transfer of training to performance presented the greatest challenge to the study.

Straus concluded by identifying some ways to improve the assessment of such training. Online data collection—for example, using mobile devices—could facilitate analysis and feedback. Better training of instructors in rating team performance could improve the reliability of those ratings. The students' work supervisors could be better held accountable to provide feedback on posttraining behavioral change. Longer-term impacts could be assessed through measures of the retention of course knowledge and attitudes, 360-degree feedback, measures of the performance of graduates' teams, and—what Straus called her "holy grail"—randomized controlled trials.

AVAILABLE MEASUREMENT TOOLS

During the discussion period, Kantrowitz and Straus elaborated on their use of the Job Adaptability Inventory and other measures of adaptability. The JAI is a proprietary tool, Kantrowitz pointed out, but other measures are available from public sources and through psychometric testing providers. And because assessment science has progressed in recent years, better tools are available, such as an exercise that puts people in simulated situations to evaluate the extent to which they are able to adapt in desirable ways. Such tools, which are available online, can measure both whether people are predisposed to react in certain ways and whether they have the knowledge and skills to do jobs that require adaptability. Straus added that many of these tools are essentially generic, so they can apply to any educational context, not just a particular course for army officers.

But, Straus cautioned, this work is in its early stages and best practices or best instruments still do not exist. Kantrowitz concurred, noting that good work is being done "in pockets" but that interest in adaptability has not caught on in many companies and industries. "It is not deeply entrenched in a lot of companies' talent management programs, which is too bad, because it is clearly [a] pressing" need. Nick Donofrio observed that many companies have management programs, but they are mostly for specific purposes and have not been publicized.

Straus noted that much of this work on evaluating adaptability has been done in assessment centers, which offer a strong approach that has been used in many domains. "But," she added, "it is also a costly approach, [which] is one of the reasons it is not more prevalent." Perhaps the advent of computerized tools will make the assessment of adaptability less costly and more prevalent, she said. Joppa observed that much of this assessment work has been based on trial and error, since not enough is yet known about the role of adaptability in moving an employee from one job to another within a company.

6

Training and Organizational Change

BOX 6-1
Highlights of Panel Presentations

- A well-defined career framework with specific work elements can guide the acquisition of multiple skills, allowing employees to build a career path while making the company more nimble. (Jim Johnson)
- Partnerships between companies and community colleges can build skills among local students, using nontraditional methods to develop problem-solving skills. (Blake Consdorf)
- Companies are less likely to pay for their employees to develop adaptability skills such as communication, program solving, and critical thinking than they are to pay for more narrowly focused skills, such as coding. Community colleges need to help change that dynamic. (Steve Partridge)
- College and university engineering programs should expect all their students to develop an interest in both the economic and societal value they can create in their jobs. (Douglas Melton)
- Evaluation of training and educational programs can identify facilitating factors and challenges in developing and applying adaptability skills. (Carra Sims)

In two "lightning round" panels presenters briefly described activities related to adaptability in their organizations. Workshop participants then worked in teams to identify major points, common themes, and remaining questions that emerged from the presentations.

The first lightning round on the use of training and organizational change to enhance adaptability was moderated by Mary Ann Pacelli, manager of workforce development at the Manufacturing Extension Partnership, National Institute of Standards and Technology, US Department of Commerce. (The second lightning round is reported in chapter 7.)

CAREER PATHS FOR A FLEXIBLE WORKFORCE

Jim Johnson, Jr., vice president of US production operations for S&C Electric Company, which designs and manufactures electrical equipment, began by stating that one third of the hourly workforce at S&C Electric will be eligible to retire over the next eight years. Another third has been with the company less than five years, and half of all turnover at the company comes from this second group of workers.

To provide its employees with a recognizable career path while also building a flexible workforce, the company developed what it calls an *hourly career framework*, which includes a skills and flexibility program called the *work elements program*. Rolled out among welders and then extended to all 14 functional career paths in the company, the program is based on paying workers for developing skills and flexibility. Jobs are organized into four skill levels—basic, intermediate, advanced, and complex—with multiple work elements that employees can use to enhance their skills and flexibility. Work elements are specific tasks or procedures for operating a specific piece of equipment, such as "Low Voltage Wiring Assemblies" or "High Voltage Test Cage."

As employees acquire multiple work elements at the same skill level, they can work on different product lines, which enables the company to stay nimble. And feedback from employees has shown that they value being able to create paths for themselves: they can come in with low-level skills and advance through the company. As Johnson put it, they "have more than just a job. They can have a career."

The career framework even helps employees prepare for jobs at the company that do not yet exist. In the past, S&C made electrical products that did not have embedded intelligence, but today about half of its products do, and that percentage will continue to grow, said Johnson.

As a result, the company needs fewer people to fill jobs that are mostly mechanical in nature. Instead, it needs people who can advance into jobs in which demand will grow in the future. "That's how we guide our people to naturally go where the work is going to be."

CREATING A HOMEGROWN WORKFORCE

Felsomat USA, which builds machine tools, robots, and other automation equipment for the Big Three auto companies, employs a vast array of people but was having a hard time keeping talented workers. Hundreds of thousands of high-paying jobs in manufacturing are going unfilled, said Blake Consdorf, the company's chief executive officer and president, "and it's only going to get worse with the retiring baby boomers. The opportunities are huge if we can get kids engaged."

Felsomat partnered with a local community college to teach robotics, develop and recruit the best talent, instill skills in real-world problem solving, and create the next-generation workforce. The classes at the community college do not follow a step-by-step learning process. Rather, an instructor sometimes leaves out some parts or present an incorrect drawing, challenging students to be adaptable in the face of real-world problems. Graduates of the program got a two-year degree that led to jobs at Felsomat and other companies, including those that used Felsomat's products. "They have a jumpstart in their career in the value they bring to the company," said Consdorf. "And some of the first ones we hired turned into the best engineers, programmers, and managers in the company." In addition, because the students were from the local community they did not require high recruitment and placement fees and they were more likely to stay.

Every city has programs that can help fund such programs, said Consdorf. But the approach requires an ongoing commitment. Felsomat has worked to build awareness through commercials and other forms of outreach. "Getting the kids excited was one aspect, but you have to get the parents excited, too."

Manufacturing continues to be a dirty word in America, Consdorf said, and "we have to change that mentality." Average wages in manufacturing are 20 to 30 percent higher than the national average, and the highest technologies in the world are being used in manufacturing, he said. Indeed, he suggested that a Felsomat University would be appropriate given the complexity of machinery and automation.

ALIGNING WORKFORCE TRAINING WITH WORKFORCE NEEDS

The United States does not have a systematic view of workforce development, said Steve Partridge, vice president of workforce development for Northern Virginia Community College (NVCC). It has pockets of excellence, often involving a partnership between a single college and a company, and simply hopes that these examples spread. In contrast, countries like Germany have much more fully developed systems of workforce development—although Germany, he added, is under pressure to move toward a system more like that in the United States as more students are going to university because of the need for higher-level skills.

Partridge said that his job is "to make sure every employer in Northern Virginia can find a workforce." He and his colleagues work with about 2,100 employers a year, which is just a fraction of the number of employers in the state.

But many of these employers say the same thing: their current workers are nearing retirement and young people are not well prepared to enter the workforce. "Which makes sense, if you think about it," said Partridge, "[because] just a generation ago about 70 percent of high school students left high school having had a summer job of some sort. Now less than a third do." Today's students are very busy with tutoring, sports, clubs, and other activities, but they tend not to acquire the experience with jobs that previous generations did.

Educational institutions and parents devote some attention to career awareness and preparation in the United States but not to the same extent as in other countries. Again using Germany as an example, Partridge said that 12-year-olds there are expected to arrange their own one-week internships, after which they "get some very reliable feedback about what their strengths are, and what their weaknesses are, which makes them better prepared when they're making career decisions."

An NVCC labor market research group has been analyzing skill deficits identified by employers: communication was number 1, problem solving/critical thinking number 2, and relationship management number 3. But Partridge pointed to an inverse relationship between the skills companies say their employees need and those they will pay to develop in their employees. Many companies will pay for their employees to develop coding skills, yet these are near the bottom of what they say they need. In contrast, only about 5 percent of companies will pay to develop communication, problem solving, critical thinking, or relationship management skills.

Community colleges need to help change that dynamic. "We can train for anything," said Partridge. "Our people are good, they can learn, but we don't give good market signals about what skills are needed.... If your company doesn't tell you how to invest in yourself, how do you know what skills to go out and be [learning]?"

Northern Virginia Community College is seeking to fill this gap by working with partners in the community to create a database of where the jobs are, where the jobs are going, and what skills employees need to stay competitive. America cannot afford another experience like that of the last recession, Partridge said, where many people who had not gone back to the classroom in 20 years or more lost their jobs. "How do we change that mindset?" he asked. "That's new for a college, but we're going to own that."

A NEW NORMAL FOR ENGINEERING PROGRAMS

The Kern Entrepreneurial Engineering Network (KEEN) is a partnership of higher education institutions adapting their undergraduate engineering programs by working to increase students' curiosity about the future and develop their interest in economic and societal value, said Douglas Melton, program director at the Kern Family Foundation. KEEN's 31 partner institutions represent about 58,000 undergraduate engineering students and about 3,500 engineering faculty members—about 10 percent of all undergraduate engineering students and faculty members in the country.

The partnership seeks to put value first and tools second, he said, so that "know-why" is supported with "know-how." This transformation cannot be done by simply adding one or two classes, he explained. When students start a new project, they consider the demands of stakeholders, the economics of engineering solutions, sustainability, the broader societal context, and possible unintended consequences.

The key statement of the partnership is that the adaptability of the future technical workforce depends directly on the adaptability of its education. "You have to get that in place." And the Kern Family Foundation is considering strengths, weaknesses, opportunities, and threats (SWOT) analysis of adaptability to speed progress.

Melton quoted Klaus Schwab, founder of the World Economic Forum: "In the new world, it's not the big fish that eats the small fish.

It's the fast fish that eats the slow fish."[1] The KEEN partner institutions are the fast fish of engineering, he said. While each is unique and adaptation is occurring at different rates, all are persistent and committed. The result, he predicted, will be a new normal for engineering programs.

EVALUATING AN ADAPTABILITY PROGRAM

Carra Sims, senior behavioral and social scientist with the RAND Corporation, elaborated on RAND's evaluation of the US Army Asymmetric Warfare Adaptive Leader Program (AWALP) described by Straus in the previous session (chapter 5). AWALP uses an outcomes-based approach that emphasizes flexibility in how the tasks to achieve an outcome are executed. It encourages trainees to take the initiative and adjust their actions to adapt to a situation, using independent thinking and problem solving.

The evaluation of the program was designed to promote ongoing improvement by identifying both facilitating factors and challenges in applying adaptability knowledge and skills on the job. After the training, the instructors (guides) facilitated an after-action review to promote self-discovery of lessons learned and found numerous positive outcomes and impacts of the training.

However, the training as structured is resource intensive, Sims noted. Units had to be willing to part with personnel for the ten days of the training session (without coverage of participants' job responsibilities). Some of the scenarios used paid actors, and a lot of time went into creating the structure of the course. The program also had a high ratio of guides to trainees.

The most common areas of implementation three months later were things that the trainees could control themselves, Sims reported; these included coaching, training, delegating, and seeking subordinate input. Challenges to implementation included the command climate and entrenched leadership.

Possible approaches to future training include short train-the-trainer versions of AWALP elsewhere in the army and outcomes-based training designed to enhance some adaptability dimensions even if it is not specific to adaptability.

---

[1] "Are you ready for the technological revolution?" February 19, 2015. Available at https://www.weforum.org/agenda/2015/02/are-you-ready-for-the-technological-revolution/.

TAKEAWAYS AND REMAINING QUESTIONS

After the presentations, workshop participants discussed takeaways and remaining questions.[2] Among the takeaways were the following:

- Adaptability is related to sustainability and durability, in both educational and workforce settings.
- An emphasis on mindsets in addition to skill sets has the potential to be transformative in education.
- Companies and educational institutions must keep the communication lines open about existing and anticipated company needs, especially in nontechnical skills.
- Community colleges can create excitement through competition, challenges, and partnerships.
- Workers can be educated to move up a career later rather than headed into dead-end positions.
- A healthy innovation ecosystem requires diverse inputs, and flexibility and adaptability are one way of achieving diversity.

Various participants raised the following questions:

- To what extent can best practices in building or understanding adaptability be generalized from one industry to others?
- To what extent do theories of adaptability in one domain apply in other domains?
- How can industries be convinced to invest in the workforce skills that they say they need?
- How can organizations put systematic and appropriate programs in place to address gaps in workforce development?
- Which skills can be built in a self-guided way, and which need support?
- How can experiential learning be structured so as to enhance adaptability?
- What is the concrete nature of adaptability skills? How widespread are they and how can they be measured?

---

[2] The following lists are the rapporteurs' summary of the main points made by individual speakers during the general discussion. Given the unstructured nature of the discussion, it was not possible to clearly identify every speaker or attribute every comment or idea. The statements have not been endorsed or verified by the National Academies of Sciences, Engineering, and Medicine.

- How can the measurement of outcomes linked to adaptability be improved over time?
- How can effective measures for quantifying adaptability be developed?
- How can measurement systems for adaptability be made more affordable and accessible to more organizations?
- What roles should local leaders, such as mayors and community college administrators, play in local and regional workforce development? For example, how can they help draw local employers into more active involvement in workforce preparation?
- How does adaptability differ from and relate to resilience? For example, when is the persistence of adaptations desirable?
- What are the best ways to identify and share best practices?
- Which adaptability skills are best developed in K–12 education and which in higher education or later?
- How can adaptability best be taught in the K–12 environment?
- Has anyone studied the effect of gender on adaptability? Can adaptability programs be used as a way to retain women engineers?

DISCUSSION

In closing this session, the presenters addressed a question posed by the moderator about what they saw as the most important steps to advance training and organizational change. (Additional steps, suggested by the workshop moderators and attendees, are presented in chapter 8.)

Johnson said that education and industry need to be more in sync. Are two-year and four-year institutions preparing students with the kinds of skills that companies need in the workforce? Industry needs to provide information and feedback to institutions of higher education to improve this alignment, he said.

Consdorf emphasized the broader culture. "Do we ever talk about why, or vision, or culture, in the school system?" An emphasis on the bigger picture might make it easier to explain to displaced workers why their jobs are gone and not coming back and why they need to be adaptable to respond to such changes. He also asked whether researchers could find any correlation between professions or degrees and levels of adaptability, because that information might help foster adaptability.

Partridge cited the value of internships. Most businesses do not do internships for high school students, and if they do they tend to use the internship more as a recruiting tool. Internships and job shadowing on the scale that is done in Germany, for example, are "unheard of here." For most college graduates, their first job after college is now probably their first job ever. In addition, two thirds of today's college students will work in jobs that do not yet exist, he pointed out. "How is higher education or K–12 education going to prepare students for jobs when we don't know what they are?"

Melton mentioned that companies have made changes to increase the speed, quantity, and variety of innovations they incorporate, but innovation and adaptability require systems-level thinking. "We have 31 institutions [in KEEN] that have bought in wholesale to this mission. That leaves about 390 that are working it out their own way."

Sims made the point that adaptability and innovation must be approached at multiple levels, and added that people have to be inspired and organizations need to not stifle that enthusiasm. "Societally, we need to give people the space to be creative and to change."

The panelists addressed the question about how to motivate people and institutions to take needed steps. Johnson said that the driver for his employee-owned company's program was to benefit the workers. "[W]e make decisions based on long-term investments, and it all starts with our people." Upfront costs were not so important given the expected value to the company. "Of course, we're in business to make money, don't get me wrong," he said, but the culture of the company enabled it to take a chance that the program would pay off.

Consdorf agreed that costs were associated with the program at the community college, though they were soft costs rather than the costs of providing hardware. "[Two of] our people were teaching, and for a year half their day on Tuesdays and Thursdays was dedicated to teaching those kids, not working on our machines." But the expense was essential, since "the future is dependent on it."

Partridge noted that discussions of costs need to be handled carefully, because US companies can be scared off by large projected costs. "The way it's presented sometimes can make it harder to get people to come along with you." He also pointed to a problem with the valuation of companies, because workforce talent is not valued on the bottom line of a company's books. Even if a company has the top people in the world, that talent will not necessarily be quantifiable in accounting terms.

The German program could not be moved wholesale to the United States, he acknowledged, and is changing as pressure builds to send more students to university instead of along the vocational track as workforce demands change. But apprenticeships cannot be seen as the lesser choice in the United States, he continued, because parents will not want to choose an option for their child that is perceived as less beneficial. For that reason, the United States should do a better job with the "stackability" of degrees and certificates, he said, so that students can continue their education. That is the case in Germany, where someone who goes down a vocational route can later change to a university route. In Germany "there [is] no wrong door," he said.

In the United States, Partridge said, marketing a two-year degree as "the other college degree" is a mistake. "Parents aren't stupid. They're going to say, 'That's not a college degree.'" Two-year degrees need to be able to lead to a bachelor's degree or a master's degree.

Melton reported that, rather than work with individual institutions, the Kern Family Foundation decided to work with a network. The result has been an interdependency "that comes from relationships [and] regular meetings, [which] led to an acceleration in sharing of the work."

7

K–12 Education and Out-of-School Learning

> **BOX 7-1**
> **Highlights of Panel Presentations**
>
> - An emphasis on adaptability has proven especially effective with women and underrepresented minority students as they gain confidence and experience. (Brynt Parmeter)
> - Articulation of an economic value proposition can help students determine how to find jobs that will pay them to do what they want to do. (Bernie Lynch)
> - "New-collar jobs" that require a high school diploma but not necessarily a four-year degree call for the same adaptability skills as those needed for other professions. (Grace Suh)
> - Because every job has technical components, all students need to learn to be adaptable. (David Greer)
> - After-school and summer programs provide tremendous opportunities to enhance adaptability. (Chris Smith)

The second lightning round, on the potential of K–12 education and out-of-school learning to enhance adaptability, was moderated by Betsy Brand, executive director of the American Youth Policy Forum.

FLEX FACTOR

NextFlex is a program to inspire young people to go into advanced manufacturing and higher-technology jobs in a way that resonates with them, with their parents, and with their schools. Based in Silicon Valley, it began as a small pilot with just eight students and now serves more than 2,000.

Brynt Parmeter, director of workforce development at NextFlex, spent almost 25 years in the army and noted that the business environment has some parallels with combat. Combat is an unforgiving and difficult environment. To prepare new soldiers for live fire, trainees first do a dry run of the situation, then they walk through it with blank rounds, and then they do it with live fire. All this is still not a good indicator of what combat will be like, with its many unknowns.

Flex Factor immerses high school students in the world of advanced manufacturing and entrepreneurship through project-based learning activities. Students come up with a problem, think of a product that solves the problem, form product development teams, and then pitch their ideas to a panel of judges. "By not giving them the problem, you plant deeper roots in the brain of that young person," Parmeter said.

He noted that the approach has proven especially effective with women and minority students "because they think of a problem that is important to them." They gain confidence and experience, and young women in high school serve as recruiters and mentors for the program so that participants have opportunities to see themselves in the future. "I love one of the sayings we heard," said Parmeter: "'If you can't see it, you can't be it.'"

The need for the program to generate its own revenue has advantages, he explained, "because then you listen to your customer and you'll do what both your client and customer want you to do instead of what you think is necessary." The program has been in communication with companies to export and replicate it elsewhere.

MADE RIGHT HERE

When startup companies begin to succeed in the United States, they typically move to China to begin manufacturing the product. Made Right Here is seeking to change that dynamic, said Bernie Lynch, the organization's chief executive officer.

To develop a systemic framework for startups' local and domestic manufacturing, a group of partners, with funding from the US Department of Labor's Workforce Innovation Fund, created an innovative apprenticeship model that featured front-loaded training. Breadth of knowledge was emphasized, not only depth. About 200 unemployed adults of all ages were involved in the training and were teamed up with startup founders. Training was project based, with a feedback loop to enhance learning. The learning modules were not stackable but modular, with an emphasis on iteration, innovation, and creativity.

Made Right Here asks the people with whom it works what they want to do with their life and then works with them to explore how they might achieve their goal. "What would that pathway look like, what are the steps, how do you get there?" The focus is on each individual's economic value proposition. "How can you provide an economic value proposition in a society that will pay you to do what you want to do?"

One important decision was not to use the term *apprenticeship*, said Lynch. Apprenticeships imply blue-collar jobs, while internships imply white-collar jobs. Making the products of the future requires very diverse skills, including expertise with software, hardware, creative thinking, and prototyping. To reflect these sophisticated skills, the partnership uses the term *makership*. The experiences are "not secondary work [but those of] a profession."

P-TECH SCHOOLS

In 2011, IBM started the P-TECH program (Pathways in Technology Early College High Schools) as part of its commitment to education, said Grace Suh, director of education and corporate citizenship at IBM. The program partners with K–12 schools and community colleges to build not just academic and technical skills but also professional skills, "because those are the kinds of skills that we hire for in IBM."

The company calls jobs that require a high school diploma but not necessarily a four-year college degree *new-collar jobs*. By 2024, she said, an estimated 16 million such jobs will be created in the United States, whereas from 2008 to 2016 the US economy lost 7 million jobs for those with only a high school diploma.

The P-TECH program emphasizes problem solving, critical thinking, communication, leadership, and adaptability. "Those are the kinds of skills we're trying to foster in these young people so that when they graduate they are ready for jobs." Graduates of the program have both a high school diploma and a two-year industry-recognized postsecondary degree.

P-TECH schools are open enrollment. Some students might enter reading at a fifth-grade level, but they are nevertheless expected to get a postsecondary degree. Most students are the first in their family to graduate from college and are earning their degree for free. Students receive the supports they need to get a degree, including cultural support—"at IBM, we say that culture is everything," said Suh. They also develop skills that can be applied in many different professions, so if they decide they want to be a lawyer or human resources professional, they have the skills to go in those directions.

The program annually reviews a map of academic, technical, and professional skills and compares it with the curriculum to identify gaps where the curriculum can be enriched. Students have mentors from IBM and other industry partners who model both technical and professional skills. They go to worksites, do job shadowing, and engage in a skills-based paid internship.

The program went from one school in Brooklyn in 2011 to 70 schools at the time of the workshop. IBM leads in eight of the schools; the rest are led by 400 other business partners, many working in consortiums with single schools.

Suh affirmed that "businesses want to get involved once they know how. They're looking for talent, and rather than waiting at the end of the talent pipeline they want to reach in and help develop that talent." IBM helps businesses learn about the program through meetings, online guides, and other resources so that businesses can get a high return on investment from their involvement.

The program is active in six states, soon to be eight, "and we're hoping to be in more." The program works with governors in each state to ensure the funding and policies required for long-term sustainability.

The first cohort of students had a 56 percent graduation rate, more than four times the on-time national community college graduation rate of 13 percent, Suh reported.

Graduates are the first in line for jobs at IBM, "because that's a promise we make to the young people, and it would be hollow if they weren't graduating with skills that industry actually needed." Of 100 graduates to date, IBM has hired 11, some of whom have already received promotions. "It's been an incredibly rewarding opportunity for us and one that we want to see across the United States."

PROJECT LEAD THE WAY

David Greer, senior vice president and chief programs officer for Project Lead The Way, made the case that, while not every company is a technology company, every company is a technology-enabled company and all companies need a workforce that can rise to that challenge. "We shouldn't focus just on engineers," he said. "We need to focus on digital citizens and how we can create productive workers and citizens throughout our educational programs," from prekindergarten through college and beyond. "You don't just start at the end of the pipeline; you start at the very beginning."

Project Lead The Way started two decades ago with one school and has grown to 11,000 schools in all 50 states. It has approximately 3 million students and has trained more than 55,000 teachers at all levels on problem-based, hands-on learning. Greer framed the program's motivation: "This generation wants to solve problems that will impact their community and the world. How can we tie education into that?"

The program seeks to engage students with relevant content and provides them with foundational knowledge that they need to be successful while they work on open-ended projects and problems. Exposing them to career opportunities often and at every educational level, it sets high expectations and inspires students to rise to the challenge. They collaborate, learn from each other, and dig into real-world scenarios.

Students can explore three pathways: computer science, engineering, and biomedical science, with activities at all educational levels. When students begin in kindergarten, this approach becomes the new normal for them (and for the teachers). In this way, students have a conveyor belt of opportunities throughout their schooling with support at all stages, including higher education.

The program represents a long-term investment, but early involvement is critical, since research shows that girls and minorities begin opting out of mathematics and science as early as second grade because they find these subjects "too hard." "There's no reason [they have] to be too hard. If we can provide these pathways and expectations for our students, they can succeed."

Project Lead The Way also conducts professional development for teachers, because they "make everything happen," said Greer. The teacher training emphasizes student engagement through open-ended projects and problems, flipping the usual approach of lecturing and memorization. "Can you imagine if we treated patients the same way we treated them in hospitals 100 years ago?" Greer asked. "But somehow we're educating our students the same way we educated them 100 years ago. Why is that okay? It's not, and I think we can do better."

BOSTON AFTER SCHOOL & BEYOND

Schools are critical in building the skills associated with adaptability, but school-aged children spend 80 percent of their waking hours outside of school, noted Chris Smith, executive director of Boston After School & Beyond. After-school and summer programs therefore provide a tremendous opportunity to foster adaptability, although they tend to be fragmented—Boston alone has more than 1,000 summer programs. "They have different governance models, they have different fundraising strategies, they have different measures of success," he said. "But it's still fertile ground for these skills." Boston's efforts at harnessing these programs serve as one model.

After-school and summer programs often have fewer constraints and are more flexible than school programs. And because they are voluntary, they can tap young people's intrinsic motivation, making learning and skill development immediately relevant. They also can promote diversity, of approaches as well as participants.

Boston After School & Beyond focuses on developing the skills that young people will need throughout their lives. These skills need to be taught, learned, practiced, and measured, Smith said. "You can't merely hear about them and get better at them; they require real experiences."

Summer and after-school programs can make the community a classroom for students. In Boston, programs take advantage of museums, colleges, workplaces, nature preserves, and many other settings to develop skills. In the process, program participants develop dynamic peer learning networks in which they learn from each other as well as from adults. At the same time, the programs offer an opportunity to engage with the business community to foster the skills they will need.

Effective youth development results in college and career readiness, said Smith. "You might ask, 'Aren't these programs supposed to be fun? Should they really be about workforce development? Can workforce development be fun?' We think the answer is 'yes!' Challenge and support are two sides of the same coin, as are rigor and engagement, and this is what we're working on."

One relative shortcoming observed in these programs is that they are not particularly good at leaving time for reflection. Smith also noted that he would welcome research on the transferability and portability of the skills developed in after-school programs. When you "develop them in these settings, can you apply them in another setting when it's needed?"

KEY TAKEAWAYS

Working in teams, the workshop participants articulated takeaways from the lightning round[1]:

- With more respect, attention, and resources, two-year colleges could have higher expectations for themselves and their students.
- Much learning takes place outside formal educational environments, and social support systems could strengthen such out-of-the-classroom learning.
- New terms such as *maker professionals* and *new-collar jobs* can reflect the sense of agency and identity that supports adaptability.
- Resilience and empowerment help create the confidence and flexibility that underlie adaptability.
- Sharing examples of best practices and successful activities from schools and businesses will accelerate the changes that are needed.
- A resilient system has layers of support and supporters so that no one institution or person acts as a gatekeeper.
- A better societal support system, such as a better healthcare system, could support flexibility and creativity in the workplace.
- The development of confidence can enhance an employee's ability to adapt.

In this lightning round, the panelists also reflected on what they had heard, both from each other and the various participants:

- An overemphasis on testing can teach students that the exploration and flexibility needed for adaptability are a liability.

---

[1] The following lists are the rapporteurs' summary of the main points made by individual speakers during the general discussion. Given the unstructured nature of the discussion, it was not possible to clearly identify every speaker or attribute every comment or idea. The statements have not been endorsed or verified by the National Academies of Sciences, Engineering, and Medicine.

- Giving students the opportunity to work in teams can provide them with the confidence and inspiration they need to excel as individuals.
- Public-private partnerships can break down silos and make use of differing expertise.
- Meaningful exposure to workplace settings, not just touring a plant or facility, can help students develop the skills they will need in those settings.
- No one program or approach can meet the need for adaptability, though there may be common elements to programs that are successful.

8

Possible Next Steps Suggested by Participants

In the final session, the panel moderators summarized the messages they heard during the workshop as well as participants' comments on opportunities and gaps to be addressed in efforts to move forward.[1] This chapter lists possible next steps identified by the moderators and workshop participants to address shifts in the business model and in the nature of work that call for adaptability. The steps are categorized by major stakeholders, although actions often involve collaborations across sectors. In addition, appendix D presents a categorized list of ideas, questions, and suggestions noted by workshop participants during the lightning rounds.

Individuals

- Identifying the value proposition of adaptability for individuals could clarify priority actions.
- Activities and programs to develop other skills could be designed to have a side benefit of helping individuals become more adaptable.

---

[1] The following lists are the rapporteurs' summary of the main points made by the panel moderators in the final session and by individual speakers and participants earlier in the workshop. Given the unstructured nature of the discussion, it was not possible to clearly identify every speaker or attribute every comment or idea. The statements have not been endorsed or verified by the National Academies of Sciences, Engineering, and Medicine.

Companies

- Identifying the value proposition of adaptability for companies could clarify priority actions.
- Changing the language used to describe particular types of jobs and workers to remove negative connotations would reduce barriers to adaptability associated with those jobs.
- The use of new terms such as *maker professionals* and *new-collar jobs* can reflect the sense of agency and identity that supports adaptability.
- Strong leadership can create a culture of adaptability and help employees reach their full potential.
- Sharing examples of companies' best practices and successful activities in training their employees to meet workplace changes will accelerate the changes that are needed.

Researchers and Data Analysts

- Improved, more accessible, and less expensive analytics of adaptability could drive training efforts, data gathering, and a broader adaptability agenda.
- Further study of individual, team, and organizational adaptability can identify their commonalities and differences.
- Mapping the skills required in modern economies could help align training and education with workforce needs.
- A tool that can measure adaptability across multiple sectors and regions could provide data that can lead to continuous improvements in workforce skills.

Educators

- Establishing appropriate roles for K–12 education, higher education, and out-of-school activities in fostering adaptability could benefit students at all levels.
- Partnerships among educational institutions, companies, and government can focus attention on adaptability across sectors.
- Connecting education to real-world challenges could help develop skills needed in the modern workplace.
- Sharing examples of best practices and successful activities from K–12 and higher education will accelerate the changes that are needed.

- Educational institutions at all levels need to inculcate cultures that foster change.

General

- Providing students with information about educational pathways and workforce opportunities can help them make good decisions about educational and career options.
- Champions for change can help focus the attention of policymakers, business leaders, and educators on the need for adaptability.
- Policy needs to take into account that no one program or approach can meet the need for adaptability, though there may be common elements to programs that are successful.
- A resilient training system could be designed with layers of support and supporters so that no single institution or person acts as a gatekeeper.

Nick Donofrio offered some parting words for the workshop participants: "I'm an engineer. Work equals force times distance. You can push all you want, you can sweat blood, but if there's no distance [i.e., movement], technically there's no work."

References

Alvarez K, Salas E, Garofano CM. 2004. An integrated model of training evaluation and effectiveness. Human Resource Development Review 3(4):385–416.
Argote L. 2013. Organizational Learning: Creating, Retaining and Transferring Knowledge, 2nd edition. New York: Springer.
Argote L, Epple D. 1990. Learning curves in manufacturing. Science 247(4945):920–924.
Argote L, Ingram P. 2000. Knowledge transfer: A basis for competitive advantage in firms. Organizational Behavior and Human Decision Processes 82(1):150–169.
Baard S, Rench TA, Kozlowski SWJ. 2014. Performance adaptation: A theoretical integration and review. Journal of Management 4(1):48–99.
Burke CS, Stagl KC, Salas E, Pierce L, Kendall D. 2006. Understanding team adaptation: A conceptual analysis and model. Journal of Applied Psychology 91(6):1189–1207.
Colcombe SJ, Erickson KI, Scalf PE, Kim JS, Prakash R, McAuley E, Elavsky S, Marquez DX, Hu L, Kramer AF. 2006. Aerobic exercise training increases brain volume in aging humans. Journals of Gerontology Series A: Biological Sciences and Medical Sciences 61(11):1166–1170.
DeShon RP, Kozlowski SW, Schmidt AM, Milner KR, Wiechmann D. 2004. A multiple-goal, multilevel model of feedback effects on the regulation of individual and team performance. Journal of Applied Psychology 89(6):1035–1056.
Egelman CD, Epple D, Argote L, Fuchs RH. 2017. Learning by doing in multi-product manufacturing: Variety, customizations and overlapping product generations. Management Science 63(2):405–423.
Epple D, Argote L, Murphy K. 1996. An empirical investigation of the microstructures of knowledge acquisition and transfer through learning by doing. Operations Research 44(1):77–86.
Gladwell M. 2008. Outliers: The Story of Success. New York: Little, Brown.
Hansen T. 2010. Newcomer Innovation in Work Groups: The Effect of Regulatory Fit. Doctoral dissertation, University of Pittsburgh.
Hwang E, Singh P, Argote L. 2015. Knowledge sharing in online communities: Learning to cross geographic and hierarchical boundaries. Organization Science 26(6):1593–1611.
Johansson F. 2004. The Medici Effect: What Elephants and Epidemics Can Teach Us about Innovation. Boston: Harvard Business School Press.
Johansson F. 2012. The Click Moment: Seizing Opportunity in an Unpredictable World. New York: Penguin.
Kozlowski SWJ, Gully SM, Nason ER, Smith EM. 1999. Developing adaptive teams: A theory of compilation and performance across levels and time. In: The Changing Nature of Performance: Implications for Staffing, Motivation, and Development, eds Ilgen DR, Pulakos ED. San Francisco: Jossey-Bass.

REFERENCES

Kozlowski SWJ, Chao GT, Grand JA, Braun MT, Kuljanin G. 2013. Advancing multilevel research design: Capturing the dynamics of emergence. Organizational Research Methods 16(4):581–615.

Kozlowski SWJ, Chao GT, Grand JA, Braun MT, Kuljanin G. 2016. Capturing the multilevel dynamics of emergence: Computational modeling, simulation, and virtual experience. Organizational Psychology Review 6(1):3–33.

Mueller-Hanson RA, White SS, Dorsey DW, Pulakos ED. 2005. Training Adaptable Leaders: Lessons from Research and Practice. Research Report 1844. Minneapolis: Personnel Decisions Research Institutes.

NAE [National Academy of Engineering]. 2004. The Engineer of 2020: Visions of Engineering in the New Century. Washington: The National Academies Press.

NAE. 2015. Making Value for America: Embracing the Future of Manufacturing, Technology, and Work. Washington: The National Academies Press.

Ployhart RE, Bliese PD. 2006. Individual Adaptability (I-ADAPT) Theory: Conceptualizing the antecedents, consequences, and measurement of individual differences in adaptability. In: Understanding Adaptability: A Prerequisite for Effective Performance within Complex Environments, eds Burke CS, Pierce LG, Salas E. Bingley, UK: Emerald Group Publishing Ltd, pp. 3–39.

Pulakos ED, Arad S, Donovan MA, Plamondon KE. 2000. Adaptability in the workplace: Development of a taxonomy of adaptive performance. Journal of Applied Psychology 85(4):612–624.

Pulakos ED, Schmitt N, Dorsey DW, Arad S, Borman WC, Hedge JW. 2002. Predicting adaptive performance: Further tests of a model of adaptability. Human Performance 15(4):299–323.

Pulakos ED, Mueller-Hanson R, Arad S, Moye N. 2015. Performance management can be fixed: An on-the-job experiential learning approach for complex behavior change. Industrial and Organizational Psychology 8(1):51–76.

PwC (PricewaterhouseCoopers). 2014. Adapt to Survive: How Better Alignment Between Talent and Opportunity Can Drive Economic Growth. Available at https://www.pwc.com/gx/en/hr-management-services/publications/assets/linkedin.pdf.

Ren Y, Argote L. 2011. Transactive memory systems 1985–2010: An integrative framework of key dimensions, antecedents, and consequences. Academy of Management Annals 5(1):189–229.

Ridderinkhof KR, Span MM, van der Molen MW. 2002. Perseverative behavior and adaptive control in older adults: Performance monitoring, rule induction, and set shifting. Brain and Cognition 49(3):382–401.

Schunn CD, Reder LM. 2001. Another source of individual differences: Strategy adaptivity to changing rates of success. Journal of Experimental Psychology: General 130(1):59–76.

Schunn CD, Lovet MC, Reder LM. 2001. Awareness and working memory in strategy adaptivity. Memory and Cognition 29(2):254–266.

Straus SG, Shanley MG, Sims CS, Hallmark BW, Saavedra AR, Trent S, Duggan S. 2014. Innovative Leader Development: Evaluation of US Army Asymmetric Warfare Adaptive Leader Program. RAND RR-504-A. Santa Monica: RAND.

van Vianen AE, Klehe UC, Koen J, Dries N. 2012. Career adapt-abilities scale—Netherlands form: Psychometric properties and relationships to ability, personality, and regulatory focus. Journal of Vocational Behavior 80(3):716–724.

Wegner DM. 1987. Transactive memory: A contemporary analysis of the group mind. In: Theories of Group Behavior, eds Mullen B, Goethals GR. New York: Springer-Verlag, pp. 185–208.

White SS, Mueller-Hanson RA, Dorsey DW, Pulakos ED, Wisecarver MM, Deagle EA III, Mendini KG. 2005. Developing Adaptive Proficiency in Special Forces Officers. Research Report 1831. Arlington, VA: US Army Research Institute for the Behavioral and Social Sciences.

Appendix A

Workshop Agenda

WORKSHOP ON PREPARING THE ENGINEERING
AND TECHNICAL WORKFORCE FOR ADAPTABILITY
AND RESILIENCE TO CHANGE

November 2–3, 2017
National Academy of Sciences, Room 120
2101 Constitution Avenue NW, Washington, DC

THURSDAY, NOVEMBER 2

7:00 pm   Dinner and keynote speaker
(West Court)

                          Welcome
                          Theresa Kotanchek, Evolved Analytics, LLC
                          Chair, Steering Committee on Workforce
                          Adaptability

                          Introduction
                          Nick Donofrio, IBM Corporation

                          Keynote
                          Frans Johansson, The Medici Group

FRIDAY, NOVEMBER 3

8:30 – 8:45 am   **Welcome and introduction**
                          Theresa Kotanchek
                          Al Romig, National Academy of Engineering

8:45 – 10:15 am	**Panel #1 Case for change: Why adaptability matters**
	Moderator: Wanda Reder, S&C Electric
	Why adaptability is important: Findings from the *Making Value for America* report Nick Donofrio
	View from the workforce Guy Berger, LinkedIn
	View from industry Gregory Dudkin, PPL Electric Utilities Corporation
	View from education Robert Johnson, University of Massachusetts Dartmouth
10:15 – 10:30 am	**Break**
10:30 am – 12:00 pm	**Panel #2 What does adaptability look like**
	Moderator: Ann McKenna, Arizona State University
	Individual (cognitive and motivational) Chris Schunn, University of Pittsburgh
	Interpersonal (behavioral, "soft" skills) Ernest Wilson, University of Southern California
	Team (action regulation and information processing) Steve Kozlowski, Michigan State University
	Organization (knowledge transfer, memory) Linda Argote, Carnegie Mellon University

12:00 – 12:30 pm	Lunch break
12:30 – 1:15 pm	Session: Analytics/measurement (what are analytics and how are they used to make decisions on adaptability)
	Moderator: Annette Parker, South Central College Sandee Joppa, RealTime Talent Tracy Kantrowitz, Gartner Susan Straus, RAND
1:15 – 2:45 pm	Session: Company and higher education activities—training and organizational change
	Moderator: Mary Ann Pacelli, Manufacturing Extension Partnership, NIST
	Lightning round presentations (5 minutes each) Jim Johnson, Jr., S&C Electric Blake Consdorf, Felsomat USA Steve Partridge, Northern Virginia Community College Doug Melton, Kern Family Foundation Carra Sims, RAND
	Interactive session
2:45 – 3:00 pm	Break
3:00 – 4:30 pm	Session: How K–12 education and out-of-school learning opportunities can help foster adaptability
	Moderator: Betsy Brand, American Youth Policy Forum

Lightning round presentations (5 minutes each)
Brynt Parmeter, NextFlex
Bernie Lynch, Made Right Here
Grace Suh, IBM P-TECH
David Greer, Project Lead The Way
Chris Smith, Boston After School & Beyond

Interactive session

4:30 – 4:55 pm **Plenary session: Opportunities and next steps**
Theresa Kotanchek and moderators

4:55 – 5:00 pm **Wrap-up and concluding thoughts**
Theresa Kotanchek

Appendix B

Biographies of Speakers and Committee Members

Linda Argote is the David M. and Barbara A. Kirr Professor of Organizational Behavior and Theory in the Tepper School of Business at Carnegie Mellon University, where she directs the Center for Organizational Learning, Innovation and Knowledge. Her research focuses on organizational learning, organizational memory, knowledge transfer, social networks, and group processes and performance. Her work has been published in numerous journals and her book, *Organizational Learning: Creating, Retaining and Transferring Knowledge*, was a finalist for the Terry Book Award of the Academy of Management. She completed her second term as editor in chief of *Organization Science* in 2010. The Organization and Management Theory division of the Academy of Management chose her as its Distinguished Scholar in 2012. She is a fellow of the Academy of Management, Association for Psychological Science, and INFORMS. She received a bachelor of science degree from Tulane University and a PhD in organizational psychology from the University of Michigan.

Ewa Bardasz (NAE) is president and cofounder of Zual Associates in Lubrication LLC. She also teaches an SAE continuing education class on Success Strategies for Women in Industry and Business. She was a technical fellow at the Lubrizol Corporation and engineering associate at Exxon Research & Engineering. Dr. Bardasz holds over 25 patents, has published multiple technical and scientific papers, authored chapters for technical books, and is a frequent invited speaker at conferences throughout the United States and Europe. She received SAE International's 2002 Award for Research on Automotive Lubricants and 2009 Award for Environmental Excellence in Transportation. Dr. Bardasz is a fellow of SAE International and the Society of Tribologists and Lubrication Engineers (STLE), and she was elected to the NAE in 2015 in part for her contributions to the education of engineering professionals. She obtained a PhD in chemical engineering from Case Institute of Technology.

Guy Berger is LinkedIn's chief economist and in this capacity directs research in the Economic Graph program. His team's core publication is the monthly LinkedIn Workforce Report. His research interests include skills gaps, unemployment and inactivity, regional economies, productivity growth, and the impact of technology on the global workforce. Before joining LinkedIn, he was a macroeconomist at Bank of America and the Royal Bank of Scotland. Dr. Berger completed his BA in economics and mathematics at the University of California, San Diego, and his PhD at Yale, where his dissertation focused on international and development economics.

Betsy Brand is executive director of the American Youth Policy Forum (AYPF) and specializes in comprehensive approaches to helping young people be prepared for today's careers, lifelong learning, and civic engagement. She began her education policy career as a legislative associate for the US House Committee on Education and Labor from 1977 to 1983. She subsequently served with Senator Dan Quayle as a professional staff member on the US Senate Labor and Human Resources Committee (1983–89), handling all federal education and training legislation. In 1989 Ms. Brand was appointed assistant secretary for vocational and adult education at the US Department of Education, under President George H.W. Bush, where she worked for four years. From 1993 to 1998, she operated her own consulting firm, Workforce Futures, Inc., which focused on policy and best practices affecting education and workforce preparation. She served as codirector of AYPF since 1998 and executive director since 2004, and serves on various boards, including the Latin American Youth Center and Diploma Plus. She has a BA from Dickinson College.

Blake Consdorf, president and CEO of Felsomat USA, Inc., has been working in and around the manufacturing and automation industry for the past 20+ years. He started his career at an automation company called Wes-Tech in the Chicago suburbs. Wes-Tech specialized in automated assembly and machine tool automation. From there he worked at Felsomat USA for three years before joining Ellison Technologies (later renamed Acieta), a robotic integrator that works with a number of different manufacturing-based customers. He worked there for ten years in a number of capacities, including as president, and in July 2017 returned to Felsomat USA, which specializes in gear and engine component manufacturing and automation equipment. Mr. Consdorf graduated from Purdue University with a BS in mechanical engineering.

Nicholas Donofrio (NAE) is a 44-year IBM veteran who held the coveted position of executive vice president innovation and technology and was also selected as an IBM fellow, the company's highest technical honor. He is a life fellow of the Institute for Electrical and Electronics Engineers, a fellow of the UK Royal Academy of Engineering, a member of the US National Academy of Engineering (NAE), a fellow of the American Academy of Arts and Sciences, and a member of the Connecticut Academy of Science and Engineering and New York Academy of Science. He serves on the boards of directors of Liberty Mutual, Delphi Automotive, AMD, Sproxil, Quantexa, HYPR Biometrics, the New York Genome Center, the National Association of Corporate Directors, and the Peace Tech Lab, and on the boards of trustees of Syracuse University and the MITRE Corporation. He is an executive in residence at the Columbia University School of Professional Studies. Mr. Donofrio chaired the NAE Committee on Foundational Best Practices for Making Value in America. He received his BS from Rensselaer Polytechnic Institute and MS from Syracuse University, both in electrical engineering.

Greg Dudkin, president of PPL Electric Utilities, has more than three decades of experience in telecommunications, electric, and gas utility operations. He was previously executive vice president of operations at Commonwealth Edison and senior vice president of technical operations and fulfillment at Comcast Corporation. During his years in the utility business, Mr. Dudkin served as executive vice president of Energy Delivery Operations at ComEd and as operations vice president at PECO. A registered professional engineer, he began his career as an energy applications engineer. Mr. Dudkin has a mechanical engineering degree from the University of Delaware and an MBA from Widener University.

David Greer is senior vice president and chief programs officer for Project Lead The Way (PLTW), a national nonprofit organization that provides transformative learning experiences in computer science, engineering, and biomedical science for K–12 students and teachers across the United States. He leads PLTW's curriculum, professional development, and evaluation experts whose work is at the core of PLTW's mission: empowering students to develop the in-demand knowledge and transportable skills they need to thrive in our evolving world. Mr. Greer is a nationally recognized leader in information security and digital forensics education and training. In addition to creating the Tulsa Regional STEM Alliance that united over 90 organizations around STEM talent creation and retention, he has worked in K–12 and postsecondary education with the Oklahoma Department of Career and Technology Education and the University of Tulsa's Institute for Information Security; and has briefed US senators and the FBI, IRS, and NSA on topics ranging from information security to cybercrime. He serves on the Oklahoma Governor's Science and Technology Council as well as the board of directors of the Oklahoma Space Industry Development Authority. He has a BS in management information systems from Oklahoma State University and an MS in computer science from the University of Tulsa.

Frans Johansson is founder and CEO of the Medici Group, where he advises executive leadership from some of the world's largest companies (including 30 percent of the Fortune 100), as well as startups, venture capital firms, government agencies, and universities. He has authored two books, *The Medici Effect* and *The Click Moment*, and is an international speaker on the message that diversity drives innovation. Raised in Sweden by his African-American/Cherokee mother and Swedish father, Mr. Johansson has lived all his life at the intersection. He holds a BS in environmental science from Brown University and an MBA from Harvard Business School.

Jim Johnson, Jr., is vice president, US Production Operations for S&C Electric, responsible for the fabrication, assembly, and production support organizations at manufacturing locations in Chicago, Franklin (WI), Alameda, and West Palm Beach. He is active in multiple industry groups with a focus on lean manufacturing and operational excellence. He began working at S&C while still in school in various co-op and part-time roles. After graduating, he joined S&C full-time as an engineer in the facilities engineering group where he spent 7 years. He eventually moved into operations as a supervisor in the machine shop. After several promotions taking him through various assignments in US fabrication, US assembly, and international operations in Mexico and China, Mr. Johnson settled into his current role. He graduated from the University of Illinois at Chicago with a BS in electrical engineering.

Robert E. Johnson was appointed chancellor of the University of Massachusetts Dartmouth in March 2017 and began his leadership of the 8,700-student research university on July 1, 2017. He was previously president of Becker College in Worcester. A frequent presenter and commentator on issues related to the future of work, Dr. Johnson strongly believes that college graduates must be professionally and intellectually agile to compete in a world where career mobility is the norm and individuals must continuously build their personal value. A native of Detroit, he holds a PhD in higher education administration from Touro University International, a master's degree in education administration from the University of Cincinnati, and a bachelor's degree in economics from Morehouse College.

Sandee Joppa is executive director of RealTime Talent, a public-private collaborative dedicated to improving workforce alignment across Minnesota through innovation, research, and tools. Her background includes more than 25 years of experience in human resources, including executive leadership roles as well as running her own executive coaching firm. She was vice president of human resources (chief human resources officer) for Donaldson Company, a global filtration company operating in 44 countries, for a decade. Before joining Donaldson, Ms. Joppa worked for General Mills first as a corporate recruiter, then as a human resources manager at manufacturing plants and in the sales and distribution division, as a corporate diversity manager, and finally as director of human resources for the foodservice, baking products, Yoplait, and marketing communications divisions. She serves on the Luther College Board of Regents, the Luther College Audit and Outreach and Gifts Committees, and the Make-A-Wish Minnesota Human Resources Committee. Ms. Joppa earned a bachelor's degree in English from Luther College and a master's degree in industrial relations from the University of Minnesota.

Tracy M. Kantrowitz is director of Talent Solutions at PDRI, a CEB company. She has 15 years of experience developing innovative and award-winning talent management solutions, leading teams of industrial/organizational psychology experts and multidisciplinary product development teams, designing selection programs for organizations, and conducting market and scientific research related to assessment and talent management trends. Her primary areas of expertise relate to test development, validation, research methodology, psychometrics, job analysis, performance measurement, and selection system design. Talent management products developed under her leadership have earned awards from leading industry associations and publications. She also has received multiple distinctions from the Society for Industrial and Organizational Psychology (SIOP), including the M. Scott Myers Award for applied research in the workplace, the Distinguished Early Career Contributions–Practice Award, and fellow status. Dr. Kantrowitz has published extensively in leading peer-reviewed journals and presented at national scientific and client conferences, workshops, and educational webinars on topics such as computer adaptive testing, unproctored internet testing, and mobile assessment. She is program chair for the SIOP 2018 conference, past chair of the Professional Practice Committee for SIOP, and an editorial board member of *The Industrial-Organizational Psychologist* and *Industrial-Organizational Psychology: Perspectives on Science and Practice*. She holds a PhD in industrial/organizational psychology from the Georgia Institute of Technology.

Theresa Kotanchek is chief executive officer and cofounder of Evolved Analytics LLC, a data science and system design, software, and solutions provider. Before assuming this role, she spent 23 years in executive and leadership positions at Dow Chemical, including most recently as the vice president for sustainable technologies and innovation sourcing and as the chief technology officer of Dow Chemical China Company Limited. Dr. Kotanchek served as the industrial lead and working group cochair of President Obama's Advanced Manufacturing Partnership Initiative (2011–12) and was a member of the NAE Committee on Foundational Best Practices for Making Value in America. She holds a doctorate in materials science, a master's in ceramic science, and a BS in ceramic science and engineering from the Pennsylvania State University.

Steve W.J. Kozlowski, professor of organizational psychology at Michigan State University, is a recognized authority in the areas of multilevel theory; team leadership and team effectiveness; and learning, training, and adaptation. The goal of his programmatic research is to generate actionable theory, research-based principles, and deployable tools to develop adaptive individuals, teams, and organizations. He has produced over 500 articles, books, chapters, reports, and presentations. Dr. Kozlowski received the SIOP Distinguished Scientific Contributions Award and the INGRoup McGrath Award for Lifetime Achievement in the Study of Groups. He is series editor for the *Oxford Series on Organizational Psychology and Behavior* and former editor in chief and associate editor for the *Journal of Applied Psychology*. He is an editorial board member for the *Academy of Management Review*, the *Journal of Management*, and *Leadership Quarterly*, and served on the editorial boards of the *Academy of Management Journal, Human Factors*, the *Journal of Applied Psychology*, and *Organizational Behavior and Human Decision Processes*. He is a fellow of the American Psychological Association, Association for Psychological Science, International Association for Applied Psychology, and Society for Industrial and Organizational Psychology (SIOP). He was president of SIOP (2015–16) and is the SIOP research and science officer (2017–20). He received his BA in psychology from the University of Rhode Island, and his MS and PhD degrees in organizational psychology from the Pennsylvania State University.

Bernie Lynch is chief executive officer of Made Right Here, a national nonprofit providing training and education focused on developing the systemic framework for startup local and domestic manufacturing. In this role, she led a team engaged in an innovative approach to apprenticeship in the "New App for Making It in America" with the US DOL Workforce Innovation Fund. She serves as an apprentice leader and a subject matter expert in manufacturing apprenticeship with the US DOL. Made Right Here's Maker Professional Registered Apprenticeship focuses on emerging technology from ideation to commercialization, requiring agility in materials, equipment, methods, and practices, with the ability to design, iterate, innovate, and commiserate. In 2013 she codeveloped a workshop briefing with OSTP on startup manufacturing for White House staff and policymakers, leading to nationalizing the effort. She previously worked in the office of the mayor, City of Pittsburgh, as director of grants and development, launching initiatives like the Pittsburgh Promise, the Redd Up Campaign, and Authorities, Boards, Commissions, and Departments Grants Group. She serves on the Pennsylvania Energy Development Authority and is developing an additive manufacturing technician apprenticeship.

Ann F. McKenna is professor of engineering and director of the Polytechnic School in the Ira A. Fulton Schools of Engineering at Arizona State University (ASU). Prior to joining ASU she was a program director at the National Science Foundation in the Division of Undergraduate Education, and was on the faculty in the Department of Mechanical Engineering and Segal Design Institute at Northwestern University. Dr. McKenna's research focuses on the role of adaptive expertise in design and innovation, entrepreneurial thinking, and mentorship approaches of engineering faculty. She works across the disciplinary lines of engineering, education, and design and has published in diverse disciplinary venues including *Science*, the *Journal of Engineering Education*, *IEEE Computer*, *ASME Journal of Mechanical Design*, and *Teaching in Higher Education*. She received her BS and MS degrees in mechanical engineering from Drexel University and PhD from the University of California, Berkeley.

Douglas Melton is a program director for the Kern Family Foundation and works closely with the Kern Entrepreneurial Engineering Network (KEEN) of 31 partner institutions that are developing educational experiences to foster an entrepreneurial mindset in their undergraduate engineering students. Now in its twelfth year, the growing program has national aim. Dr. Melton was a faculty member for 17 years (1995–2012) in the Department of Electrical and Computer Engineering at Kettering University, where he was also program director for entrepreneurship across the university. He was previously director of research and development for Digisonix Inc. His disciplinary specializations include signal processing, acoustics, and wireless communications. He received his BS from Wichita State University, MS from the Ohio State University, and PhD from the University of Wisconsin–Madison, all in electrical and electronics engineering.

Mary Ann Pacelli is manager, Workforce Development at the Manufacturing Extension Partnership (MEP), a division of NIST in the US Department of Commerce. Her work includes advocating for manufacturing workforce priorities with related federal agencies and providing technical support to the network of MEP centers across the country for workforce-related activities. Previously she was assistant director, workforce and talent development, at MAGNET (Manufacturing Advocacy and Growth Network), an Ohio MEP affiliate center. Ms. Pacelli coordinated efforts for new business development and the delivery of training and consulting services to area manufacturers designed to improve performance of the organization. In addition, she directed activities to access federal, regional, and local funding for projects related to developing a stronger manufacturing workforce pipeline in Ohio. Ms. Pacelli has a BS/BA in marketing from John Carroll University and a master's in education in adult learning and development from Cleveland State University.

Annette Parker, president of South Central College, has more than 35 years of manufacturing industry and workforce education experience with General Motors, Lansing Community College, and the Kentucky Community and Technical College System. She served on President Obama's Advanced Manufacturing Partnership (AMP) Steering Committee 2.0, the American Association of Community Colleges Board of Directors, and the National Academies Committee on the Study of Education, Training, and Certification Pathways to a Skilled Technical US Workforce, and currently serves on the Manufacturing Institute Education Council. She holds a PhD in educational leadership from Western Kentucky University and bachelor's and master's degrees in technical education from Ferris State University.

Brynt Parmeter has a diverse background in both the public and private sectors. Before his current role as director, Workforce Development, at NextFlex, a national advanced manufacturing consortium, he was a science and technology policy fellow for the Department of Energy's Advanced Manufacturing Office; cofounder of WorkScouts, a technology platform designed to connect transitioning service members and veterans with education and employment opportunities in advanced manufacturing; and a partner in BMNT, a Palo Alto–based consulting firm focused on aligning lean startup principles to solve national security problems. Before joining the private sector, Mr. Parmeter served nearly 25 years as an infantry officer in the US Army, rising to the rank of colonel. During his time in uniform, he served in a variety of leadership, operations, and training roles, with three combat tours in Iraq between 2004 and 2009. He holds a bachelor's degree in systems engineering from the US Military Academy, several master's degrees from Louisiana State University and the US Army War College, and a graduate-level certificate in business and entrepreneurship from Stanford University.

Steve Partridge is vice president of workforce development at Northern Virginia Community College (NOVA) and a nationally recognized workforce development expert. He provides proactive support to regional economic development initiatives and strategic, collegewide leadership for the development and delivery of workforce training and its integration with NOVA's academic offerings. Mr. Partridge was previously president and CEO of Charlotte Works (NC), where he transformed the city's Workforce Development Board into a demand-driven agency assisting more than 1,300 businesses and hosting more than 125,000 jobseeker visits. He became adept at understanding employer needs and economy-driving public- and private-sector demands, making Charlotte Works a national benchmark for workforce reform. He also created educational partnerships from elementary through college institutions and formed a volunteer corps that saved employers more than $100,000 a year. He holds a bachelor's degree from the University of Arizona and a master's in public administration from Arizona State.

Wanda Reder (NAE) is chief strategy officer at S&C Electric Company in Chicago and vice chair of the US Department of Energy's Electricity Advisory Committee. After joining S&C Electric in 2004 as vice president of power systems services, she grew the service business, expanding field service and project-related work globally. Prior to S&C, Ms. Reder held leadership positions at Exelon and Northern States Power, with responsibility for asset investment strategy, standards, engineering, systems planning, and reliability and work management. She was elected to the NAE in 2016 for her leadership in electric power delivery and workforce development. As president of the IEEE Power & Energy Society, she was involved in activities to help engineering students and early-career engineers, including shepherding the launch of the IEEE PES Scholarship Plus Initiative to attract the best and brightest young engineers into the power industry. Ms. Reder has a BS in engineering from South Dakota State University and an MBA from the University of St. Thomas.

Christian Schunn is a senior scientist at the Learning Research and Development Center; professor of psychology, learning sciences and policy, and intelligent systems; and codirector of the Institute for Learning, all at the University of Pittsburgh. The Institute for Learning provides training and guidance around instructional reform to large US school districts. Over the last three decades Dr. Schunn has led many research and design projects in STEM and writing education. His current research interests include STEM reasoning (particularly the study of practicing scientists and engineers) and learning (developing and studying integrations of science and engineering or science and math), neuroscience of complex learning (in science and math), peer interaction and instruction (especially for writing instruction), and engagement and learning (especially in science). He is a fellow of AAAS, APA, and APS and a fellow and executive member of the International Society for Design & Development in Education. He served on the NAE committees on Understanding and Improving K–12 Engineering Education in the United States and on K–12 Engineering Education Standards. Dr. Schunn has a PhD and an MS from Carnegie Mellon University and a BS from McGill University, all in psychology.

Carra S. Sims is a senior behavioral and social scientist at the RAND Corporation, working primarily in Project AIR FORCE and Arroyo Center on US Air Force and US Army manpower and wellness. Her research includes investigating the effects of workplace stressors (including military deployment, discrimination, and sexual harassment) on job attitudes and performance; assessing quality-of-life issues faced by wounded warriors, military families, and caregivers; evaluating employee selection systems, including the use of the Air Force Officer Qualifying Test in the AF officer selection system; exploring the effects of organizational climate and culture on job attitudes and behaviors, and how best to change organizational climate and culture; and determining the knowledge, skills, and abilities required for success in various military and civilian jobs. Dr. Sims has a PhD in industrial/organizational psychology from the University of Illinois at Urbana-Champaign.

Chris Smith is executive director of Boston After School & Beyond. Over the past two decades, he has created, scaled, and led cross-sector partnerships in education and workforce development. Under his leadership, Boston After School & Beyond has developed a nationally recognized model of summer learning that improves student outcomes; built a citywide program performance measurement system; and cultivated a network of 180 programs serving nearly 15,000 students. He previously worked at the Boston Private Industry Council, where he collaborated with business leaders to integrate work and learning in order to help thousands of students graduate, with universities on Boston's first-ever study of college graduation rates of Boston Public School students, and with legislative leaders to address the dropout rate in Massachusetts. Mr. Smith began his career at the US Department of Education, where he coordinated partnerships for the secretary. A native of Worcester, MA, he earned a BA in American studies from Trinity College in Hartford and an MBA from Babson College.

Susan G. Straus is a senior behavioral scientist at RAND. Her research focuses on social impacts of information and communication technologies for individuals and groups in organizations. She has been a principal investigator on numerous studies of technologies for training and other aspects of education and technology use in diverse domains. Current and recent projects have addressed adaptive leader education, development of critical thinking skills, virtual technologies for collective military training, and effects of participating in distributed communities of practice for STEM education. She has published her work in numerous RAND reports and in the *Journal of Applied Psychology*, *Organizational Behavior and Human Decision Processes*, *Small Group Research*, *Group Dynamics: Theory, Research, and Practice*, and *Computer-Supported Cooperative Work*, and recently served as an associate editor of *Group Dynamics: Theory, Research, and Practice*. Before joining RAND in 2001, Dr. Straus was on the faculty at Carnegie Mellon University as an assistant professor of organizational behavior at the Tepper School of Business and adjunct associate professor of human-computer interaction in the School of Computer Science. She received her PhD and MA in industrial/organizational psychology from the University of Illinois at Urbana-Champaign and her BA in psychology from the University of Michigan.

Grace Suh is director, Education, Corporate Citizenship at the IBM Corporation, managing IBM's global education portfolio of STEM and teacher professional development programs, including P-TECH. She previously worked at the Children's Defense Fund, where she focused on child welfare policy. In addition to the corporate and nonprofit sectors, Ms. Suh has worked on education and children's issues in state and city governments and serves on a number of education committees and boards, including the Cahn Fellows Programs and Schools That Can. She has a master's degree in public policy from the John F. Kennedy School of Government at Harvard University and a bachelor's degree from Columbia University.

Ernest James Wilson III is founding director of the University of Southern California's Center for Third Space Thinking, devoted to research, teaching, and executive education on soft skills in the digital age. His most recent research focuses on critical workforce competencies and talent and skills development in the 21st century, and as a fellow at the Center for Advanced Study in the Behavioral Sciences at Stanford University, he is writing a book on using competencies via the framework of Third Space Thinking. He served two terms (2007–17) as dean of USC's Annenberg School for Communication and Journalism. He has participated in NAE panels on soft skills, was a member of the National Academies' Computer Science and Telecommunications Board, and served on the board of the Corporation for Public Broadcasting from 2000 to 2010, the last year as chair. He is a member of the American Academy of Arts and Sciences. Dr. Wilson earned a PhD and an MA in political science from the University of California, Berkeley, and a BA from Harvard College.

Appendix C

Participants List

Amy Adler
Walter Reed Army Institute of Research

Ashok Agrawal
American Society for Engineering Education

John Alic
Consultant

Kathy Amoroso
The United States Conference of Mayors

Linda Argote
Tepper School of Business, Carnegie Mellon University

Ewa Bardasz
ZUAL Associates in Lubrication LLC

Johnny Barnes
STEM Premier

Guy Berger
LinkedIn

Nikki Blacksmith
US Army Research Institute for Behavioral and Social Sciences

M. Brian Blake
Drexel University

William Bonvillian
Massachusetts Institute of Technology

Anna Bourne
EY

Betsy Brand
American Youth Policy Forum

Albert Bunshaft
Dassault Systèmes

Kapil Chalil Madathil
Clemson University

APPENDIX C

Kevin Christian
American Association for Community Colleges

Blake Consdorf
Felsomat USA, Inc.

John J. Donleavy
Utilligent

Nicholas Donofrio
National Academy of Engineering

Richard Donovan
University of California, Irvine

Greg Dudkin
PPL Electric Utilities

Ana Ferreras
National Academy of Sciences

Brian Flynn
Center for the Study of Traumatic Stress, Uniformed Services University of the Health Sciences

Drew Glassford
Boy Scouts of America National Foundation

James Grand
University of Maryland, Department of Psychology

David Greer
Project Lead The Way

David Hardy
US Department of Energy, Advanced Manufacturing Office

Dan Higgins
EY

Jonathan Hill
Seidenberg School of Computer Science and Information Systems, Pace University

Kenan Jarboe
National Academy of Engineering

Frans Johansson
The Medici Group

James Johnson, Jr.
S&C Electric Co.

Robert Johnson
University of Massachusetts Dartmouth

Sandee Joppa
RealTime Talent

Megan Judge
The United States Conference of Mayors

Tracy Kantrowitz
PDRI

Suleyman Karabuk
GE Global Research

Brian Keech
Drexel University

Melissa Klembara
US Department of Energy, Advanced Manufacturing Office

Theresa Kotanchek
Evolved Analytics LLC

Steve Kozlowski
Michigan State University

Eric Lesser
KPMG

Valri Lightner
US Department of Energy, Advanced Manufacturing Office

Bernie Lynch
Made Right Here

David Manheim
RAND Corporation

Brad Markell
AFL-CIO Industrial Union Council

Ann McKenna
Arizona State University

Douglas Melton
Kern Family Foundation

Pablo Molina
Drexel University

Rebecca Morgan
Fulcrum ConsultingWorks Inc.

Steve Olson
National Academies of Sciences, Engineering, and Medicine

Mary Ann Pacelli
National Institutes of Standards and Technology Manufacturing Extension Partnership

Annette Parker
South Central College

Brynt Parmeter
NextFlex

Steven Partridge
Northern Virginia Community College

Wanda Reder
S&C Electric Company

Al Romig
National Academy of Engineering

David Rosowsky
University of Vermont

David Ryan
Navigo

Deth Sao
Seidenberg School of Computer Science and Information Systems, Pace University

David Sarnoff
Sarnoff Group LLC

Aleister Saunders
Drexel University

Christian Schunn
University of Pittsburgh

Carra Sims
RAND Corporation

Carolyn Slaski
EY

Chris Smith
Boston After School & Beyond

Nebiat Solomon
US Department of Energy, Advanced Manufacturing Office

Andrew Steigerwald
US Department of Energy, Advanced Manufacturing Office

Susan Straus
RAND

Grace Suh
IBM Corporation

Vicki Thompson
America Makes

Sharon Tracy
Steelcase

Robin Utz
US Department of Education

Kirste Webb
Visionary Center for Sustainable Communities

Ernest Wilson
University of Southern California

Peter Winter
US Department of Energy, Advanced Manufacturing Office

Appendix D

Suggestions from Interactive Sessions

The lightning rounds included an opportunity for workshop participants to discuss points raised by the presenters and write down ideas, questions, and suggestions for change on index cards. The statements have not been endorsed or verified by the National Academies of Sciences, Engineering, and Medicine.

These notes are categorized as follows.

THE NATURE AND EVALUATION OF ADAPTABILITY

- There is still no one definition of adaptability.
- Desired learning outcomes include communication, informational literacy, creative and flexible thinking, ethical reasoning, self-direction, technology use, global orientation, leadership, professional practice, responsible citizenship, research, and scholarship. Outcomes should be communicated early and measured before and after training.
- What are transferable adaptability skills? People can adapt in one setting but when they go to another may not be able to adapt.
- What is the concrete nature of skills? Are there hundreds or thousands? Can we make progress without better conceptual understanding?
- Adaptability necessitates cross-disciplinary knowledge and a collective orientation.
- What roles do gender and confidence play in adaptability?
- Agency, identity, empowerment, and self-reliance are keys to adaptability.

- Execution does not need to be perfect. Try and learn, then try again. Adaptability is a continuous mindset, not a starting and ending point.
- More work is needed on special or target groups, including veterans, youth (especially foster care youth), and people with disabilities.
- What are the qualities of a program that builds adaptability?
- Whose measurement tools and what metrics do we use to make a compelling case for adaptability?
- How can measuring the impact of adaptability be made affordable, accessible, and generalizable?
- Can RAND's assessments be systemized and made available as a set of tools?
- Can best practices be translated into bottom-up change?
- What is the neuroscience of adaptability?

EDUCATION AND TRAINING

- How should universities incorporate "learning by doing" in a cross-section of arenas to prepare graduates for the future?
- How do we balance, combine, and locate learning of core lifelong skills (such as communication, interpersonal competency, and adaptability) with technical skills?
- How can curricula be changed to enable workforce adaptability and be aligned with industry needs?
- How can experiential learning emphasize adaptability?
- Expectations are too low for students with associate's degrees.
- Everyone needs experiential learning.
- What type of project-based learning or experiential learning is most effective?
- Educational institutions are not stuck in old ways, but their innovations need to be marketed.
- What are universities doing to help students define their passions? How does this impact their education plans?
- How can opportunities to adapt be introduced in classrooms of 50+?
- Faculty need to be more in touch with experiential learning, even though institutions can be hard to move.
- Children are taught at an early age that failure is not okay, which inhibits creativity and innovation.

WORKPLACE LEARNING

- How do you teach and promote innovation in organizations?
- Workshops could provide general training and specific problems to create empathy, cross-functional collaboration, hands-on prototyping, and a common language.
- Mapping skill sets can provide a foundation for employees to aspire and do career planning.
- Small and medium-sized employers may not have the funds for adaptability.
- How do you distinguish between people who are agile and those who specialize in "nondurable" skills?
- An individual can be adaptive, but an organization's structure may not support it.
- How do we elevate the importance of teaching soft skills so that employers will pay for them?
- If people are selected on their basis to adapt, this will inevitably leave some people out who do not fit.
- Where does adaptability fit in with a nation of improvisers?
- With increasing time pressure, how do we provide time and space to learn continuously?
- Adaptability means that the workplace and education will be synonymous to be sure company needs are met.
- How do you work with the current workforce to "reinspire" for purpose and passion?
- How does the pressure to change and the stress on people affect innovation and adaptability?
- Are there data demonstrating the hypothesized "crash" for people who specialize in skills?
- We need to understand the models to enhance workforce development strategies among Manufacturing USA institutes.
- Studies are needed on how existing workers adapt to disruptive midcareer changes.
- Further research is needed on resilience. Can you persist after change?

BUILDING SUPPORT FOR ADAPTABILITY

- The United States Conference of Mayors provides opportunities to change policies, focus on issues, reach out, and provide rapid information.
- Databases of successful experiences need to be built and shared so that everyone can implement these practices and add to the databases.
- With a diverse, disparate system, we need to develop more partnerships and work together.
- Pockets of good things exist, but they need to be shared so more people know about them.
- Flexibility and adaptability create diversity and a healthy innovation ecosystem.
- Where are the millennials at this meeting? All the speakers and participants are middle aged and older.
- More adaptable citizens need to be the new normal. This requires a long-term strategy and a culture of adaptability.
- Social support is needed for learning outside of formal learning.